Collection and Handling of Laboratory Specimens

Edited by

Jean M. Slockbower, Ph.D.

Consultant, Department of Laboratory Medicine,
Mayo Clinic, Rochester, Minnesota

Thomas A. Blumenfeld, M.D.

Vice-President of Medical Affairs
The Presbyterian Hospital in the City
of New York
Columbia–Presbyterian Medical Center
Associate Professor of Clinical Pathology
Columbia University College of
Physicians and Surgeons
New York, New York

With 22 Contributors

Collection and Handling of Laboratory Specimens

A Practical Guide

J.B. LIPPINCOTT COMPANY

Philadelphia
London St. Louis
Mexico City São Paulo
New York Sydney

The authors and publisher have exerted every effort to ensure that drug selection and dosage set forth in this text are in accord with current recommendations and practice at the time of publication. However, in view of ongoing research, changes in government regulations, and the constant flow of information relating to drug therapy and drug reactions, the reader is urged to check the package insert for each drug for any change in indications and dosage and for added warnings and precautions. This is particularly important when the recommended agent is a new or infrequently employed drug.

1 3 5 6 4 2

Library of Congress Cataloging in Publication Data
Main entry under title:

Collection and handling of laboratory specimens.

Bibliography.
Includes index.
1. Blood–Examination. 2. Diagnostic specimens. 3. Veins–Puncture.
I. Slockbower, Jean. II. Blumenfeld, Thomas. [DNLM: 1. Specimen handling. QY 25 G694]
RB45.C64 1983 616.07'56 82-14814
ISBN 0-397-50520-5

Acquisitions Editor: Lisa A. Biello
Sponsoring Editor: Sanford J. Robinson
Manuscript Editor: Leslie E. Hoeltzel
Indexer: Eleanor Kuljian
Art Director: Maria S. Karkucinski
Designer: Ronald Dorfman
Production Supervisor: N. Carol Kerr
Production Assistant: Charlene Catlett Squibb
Compositor: Ruttle, Shaw & Wetherill, Inc.
Printer/Binder: R. R. Donnelley & Sons Company

Contributors

Thomas A. Blumenfeld, M.D.
Vice-President of Medical Affairs
The Presbyterian Hospital in the City of New York
Columbia–Presbyterian Medical Center
Associate Professor of Clinical Pathology
Columbia University College of Physicians and Surgeons
New York, New York

Howard S. Cheskin, PH.D.
Department of Pathology
Columbia University College of Physicians and Surgeons
New York, New York

Marjorie Gamm
Desk Supervisor
Hilton Building
Department of Laboratory Medicine
Mayo Clinic
Rochester, Minnesota

M. Marsha Hall, M.T. (A.S.C.P.)
Administrative Assistant
Mayo Medical School
Formerly Supervisor, Bacteriology Section of Clinical Microbiology
Department of Laboratory Medicine
Mayo Clinic
Rochester, Minnesota

James D. Jones, PH.D.
Consultant in Clinical Chemistry
Department of Laboratory Medicine
Mayo Clinic
Rochester, Minnesota

Robert I, Kalish, M.D., PH.D.
Department of Pathology
Columbia University College of Physicians and Surgeons
New York, New York

Jean Koebke
Supervisor of Venipuncture Service
Methodist Hospital
Department of Laboratory Medicine
Mayo Clinic
Rochester, Minnesota

Ruth Mangan, M.T. (A.S.C.P.)
Laboratory Technologist
Mayo Medical Laboratory
Mayo Clinic
Rochester, Minnesota

Fuad K. Mansour, B.S. (N.R.C.C.)
Supervisor of Central Processing
Department of Laboratory Medicine
Mayo Clinic
Rochester, Minnesota

Eppie McFarland
Training Coordinator of Venipuncture Services
Department of Laboratory Medicine
Mayo Clinic
Rochester, Minnesota

Mary Mein
Supervisor of Venipuncture Service
Saint Mary's Hospital
Department of Laboratory Medicine
Mayo Clinic
Rochester, Minnesota

Mary Morris, C.L.T. (H.E.W.)
Supervisor
Laboratory Hematology
Department of Laboratory Medicine
Mayo Clinic
Rochester, Minnesota

Michael B. O'Sullivan, M.D.
Chairman
Department of Laboratory Medicine
Mayo Clinic
Rochester, Minnesota

Alvaro E. Pertuz, M.SC.
Section of Systems and Procedures
Mayo Clinic
Rochester, Minnesota

Robert V. Pierre, M.D.
Section Head of Hematology
Department of Laboratory Medicine
Mayo Clinic
Rochester, Minnesota

Jean Simindinger, R.N.
Department of Ophthalmology
National Institutes of Health
Bethesda, Maryland

Jean M. Slockbower, PH.D.
Consultant
Department of Laboratory Medicine
Mayo Clinic
Rochester, Minnesota

Cheryl L. Sonnenberg, M.T., B.B. (A.S.C.P.)
Blood Bank
Department of Laboratory Medicine
Rochester, Minnesota

David Sperling, B.B.A.
Administrative Assistant
Regional Laboratory
Department of Laboratory Medicine
Mayo Clinic
Rochester, Minnesota

Susan Stumpf, M.T. (A.S.C.P.)
Department of Veterinary Science
University of Minnesota, Saint Paul Campus
Minneapolis, Minnesota

Betty Winkler
Supervisor of Venipuncture Services
Department of Laboratory Medicine
Mayo Clinic
Rochester, Minnesota

Gerald Wollner, B.A.
Administrative Assistant
Community Medicine Building
Mayo Clinic
Rochester, Minnesota

Preface

An unprecedented expansion of scientific knowledge and technology in medicine has taken place since the late 1940s. Nowhere has this technological spin-off of medical research been more apparent than in the clinical laboratory. As a result, the laboratory, always an essential component of the medical-care system, has become a new focal point in patient care. In addition to meeting the challenge of increased testing volumes and the development of new and better test systems, the laboratories must be committed to quality of service. Nowhere does this quality have more importance than in the beginning of the laboratory system itself, namely, the collection of the patient specimen. The adequate performance of specimen collection requires proper training in techniques; uniform procedural guidelines; and a sincere and concerned interest in the welfare of the patient. We hope that this practical guide will be useful to the many persons who perform, teach, or manage the collection of specimens for laboratory testing.

This practical guide on the collection of specimens is an outgrowth of two activities in which the editors have been closely involved: [1] the writing of standards on specimen collection for the National Committee for Clinical Laboratory Standards (NCCLS); and [2] the presentation of a series of phlebotomy workshops sponsored by the Mayo Medical Laboratories, Department of Laboratory Medicine, Mayo Clinic.

Jean M. Slockbower, Ph.D.

Acknowledgments

The editors are indebted to the members of the National Committee for Clinical Laboratory Standards (NCCLS) Subcommittee on Blood Collection Procedures; to our contributors; to Lisa Biello and Sanford Robinson, who encouraged the editors; and to Ms. Peggy Quandt, Dr. Slockbower's secretary, without whose invaluable help there would be no book.

Contents

Part V **Training and Administrative Considerations**

Part I

COLLECTION OF BLOOD SPECIMENS

1

Venipuncture Procedures

Jean Koebke
Eppie McFarland
Mary Mein
Betty Winkler
Jean M. Slockbower

The word *phlebotomy* is derived from the Greek root words "phlebos," meaning vein, and "tome," meaning to cut. Simply translated, phlebotomy refers to the act of cutting into a vein.

Phlebotomy is a procedure with a 3000-year history. The custom of bloodletting was practiced over the centuries to help alleviate the ills of mankind. Originating from magic and religious ceremonies, bloodletting was supposed to facilitate the release of evil spirits from elsewhere in the body. It became a method for cleansing the body of ill-defined impurities.

Hippocrates (460–377 B.C.) developed the concept that health depended on the proper balance of four body humors: blood, phlegm, yellow bile, and black bile. Thus, bloodletting became a clinical concept used to adjust one of the four body humors to proper balance.[1]

In the 12th century, barbers began to do bloodletting, the red and white barber pole a symbol of their trade. This therapeutic practice reached its high point in the United States and Europe by the end of the 18th and the beginning of the 19th centuries and then began to lose favor as a treatment for illnesses. In 1882, Oliver Wendell Holmes, in his farewell address at the Harvard Medical School, referred to bloodletting as a past "wonder-worker in disease." "The lancet," he said, "was the magician's wand of the dark ages of medicine."[2] In the early 20th century, however, bloodletting again became popular. During World War I, bloodletting became a routine treatment for soldiers gassed with phosgene–chlorine who developed severe cyanosis and dyspnea.

Today, the practice of bloodletting is used primarily for diagnostic testing rather than for treatment; however, blood removal is a recognized treatment for polycythemia and hemochromatosis. Thus phlebotomy, an ancient art, is still used in today's practice of medicine. The arena most commonly employing phlebotomy is the laboratory, for which phlebotomy facilitates collection of a venous blood specimen for analysis. The most common technique to obtain a blood specimen is venipuncture. The purpose of this chapter is to present an overview of correct procedures for obtaining blood specimens by venipuncture.

BASIC STEPS FOR DRAWING A BLOOD SPECIMEN

ACCESSION ORDER

Each request for a blood specimen must include a number to identify all paperwork and specimens associated with each patient. The information given on the blood request form should be recorded on the labels.

Essential items include the following.

- Patient's complete name from identification plate or wristband
- Identification number
- Date and time specimen was obtained

Other useful items should be considered.

- Accessioning number
- Name of physician ordering the tests
- Department or unit for which work is being done

Accessioning in a Clinic

The correct types and sizes of evacuated tubes must be selected. A label is applied to each of the needed tubes and all test references for the patient *before* the blood specimen is drawn.

Accessioning in a Hospital

This procedure may differ from the clinic procedure in only one respect: In the hospital, the tubes are labeled *after* the blood specimen has been drawn, thus eliminating possible mix-up of blood specimens and labels. Before this, the labels may be kept in a master envelope that contains the blood request form and test result forms for each patient.

IDENTIFYING THE PATIENT

The phlebotomist must ensure that the blood specimen is being drawn from the person designated on the request form. The following steps represent a suggested order for ensuring patient identification.

In an outpatient setting,

- ask the patient to state his full name, including the spelling of an unusual name.
- compare the name with that on the request form and tube labels.

In a hospital setting,

- compare your information with the patient's name and clinic number found on the door of the hospital room (Fig 1-1). When you enter the room, identify yourself to the patient, stating that you have come to draw blood for some laboratory tests.

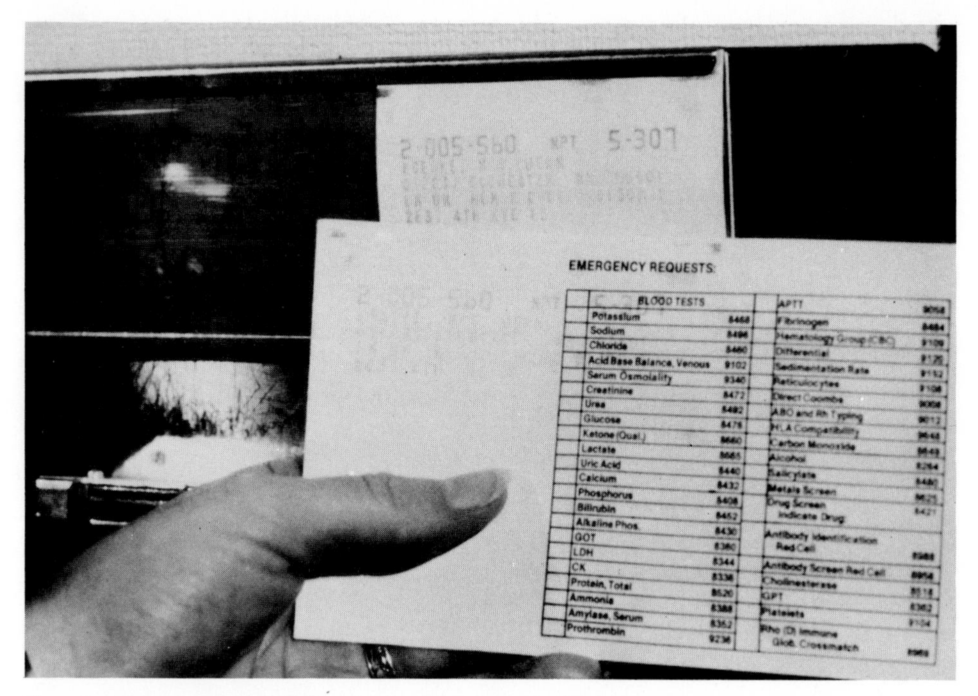

Fig. 1-1. *Comparison of information to ensure its accuracy.*

- ask the patient to state his full name, including the spelling of an unusual name (Fig. 1-2). If the patient cannot answer, ask a relative, if one is present. If not, ask a nurse to identify the patient.
- compare your information with that found on the patient's identification bracelet (Fig. 1-3).

ASCERTAINING WHETHER THE PATIENT HAS FASTED

Some tests require the patient to fast or to eliminate certain foods from the diet before the blood drawing. Time and diet restrictions vary according to the test. Such restrictions are needed to ensure accurate test results.

REASSURING THE PATIENT

The venipuncturist must gain the patient's confidence and assure him that, although the venipuncture will be slightly painful, it will be of short duration. Patients should never be told that "this will not hurt," and they should be told when the needle enters the skin so as to avoid fright.

POSITIONING THE PATIENT

Procedure for Seating the Patient

The patient should be seated comfortably in a chair and should position his arm on a slanting armrest, extending the arm so as to form a straight line from

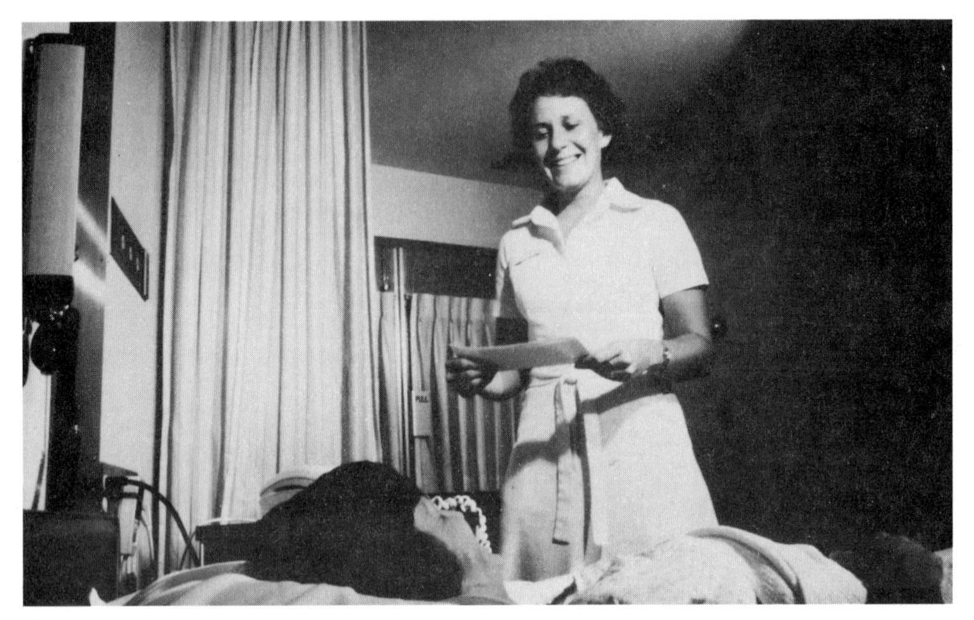

Fig. 1-2. *Ask the patient his full name.*

Fig. 1-3. *Comparison of information with that on patient's identification bracelet.*

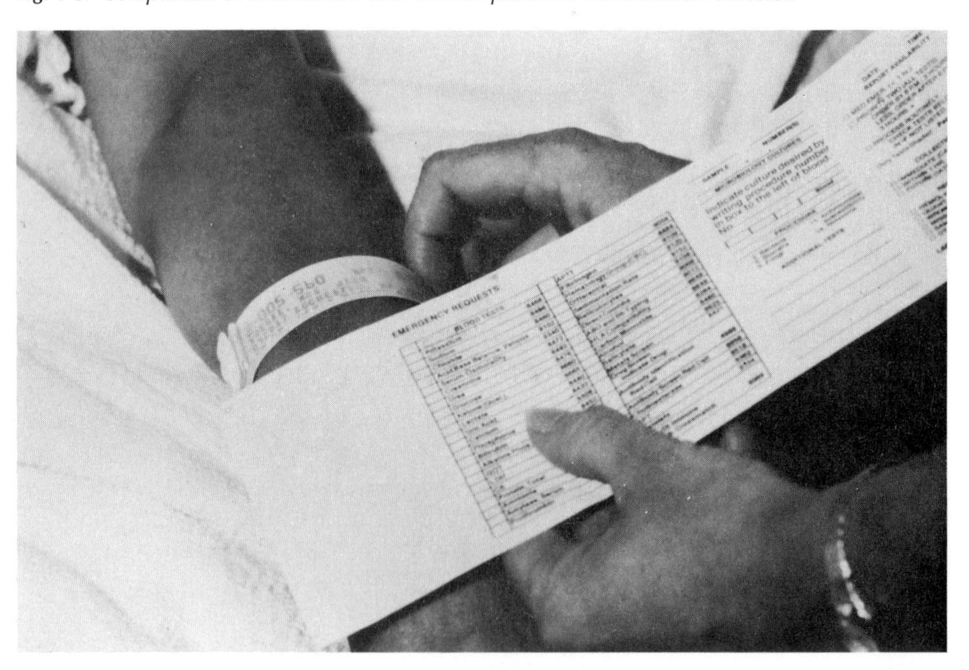

the shoulder to the wrist. The arm is supported firmly by the armrest and should not be bent at the elbow.

Procedure for Having the Patient Lie Down

The patient lies comfortably on his back. If additional support is needed, a pillow may be placed under the arm from which the specimen is to be drawn. The patient extends his arm so as to form a straight line from the shoulder to the wrist.

SUPPLIES

Assemble the following supplies.

- Collection tubes, needles, syringes
- Tourniquets
- Alcohol, 70% isopropyl, and gauze pads or alcohol prep pads (cotton balls should be used on patients who have a dermatitis problem)
- Providone–iodine swab sticks if blood culture is to be drawn
- Gauze bandage rolls

Needles

The appropriate needle is based on the patient's physical characteristics and the amount of blood to be drawn. The most commonly used bore sizes are gauges 19, 20, and 21 (Fig. 1-4)—the larger the number, the smaller the bore. If the patient has a small vein, the 21 needle is better, even though the blood will flow more slowly; small veins have a tendency to collapse if blood is drawn too quickly from them. Usually when the patient has normal veins, the 20-gauge needle is used.

Evacuated System

The evacuated system, one of two systems used to draw blood (Fig. 1-5), is the more common means of collecting specimens today. It is generally preferable to the needle and syringe because it allows the blood to pass directly from the

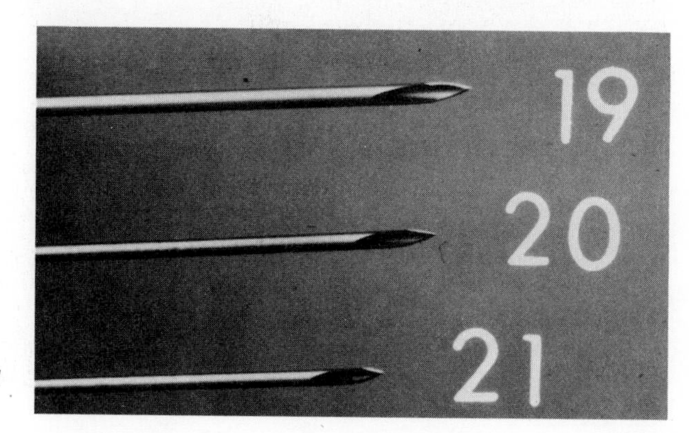

Fig. 1-4. *Gauges 19, 20, and 21 needles.*

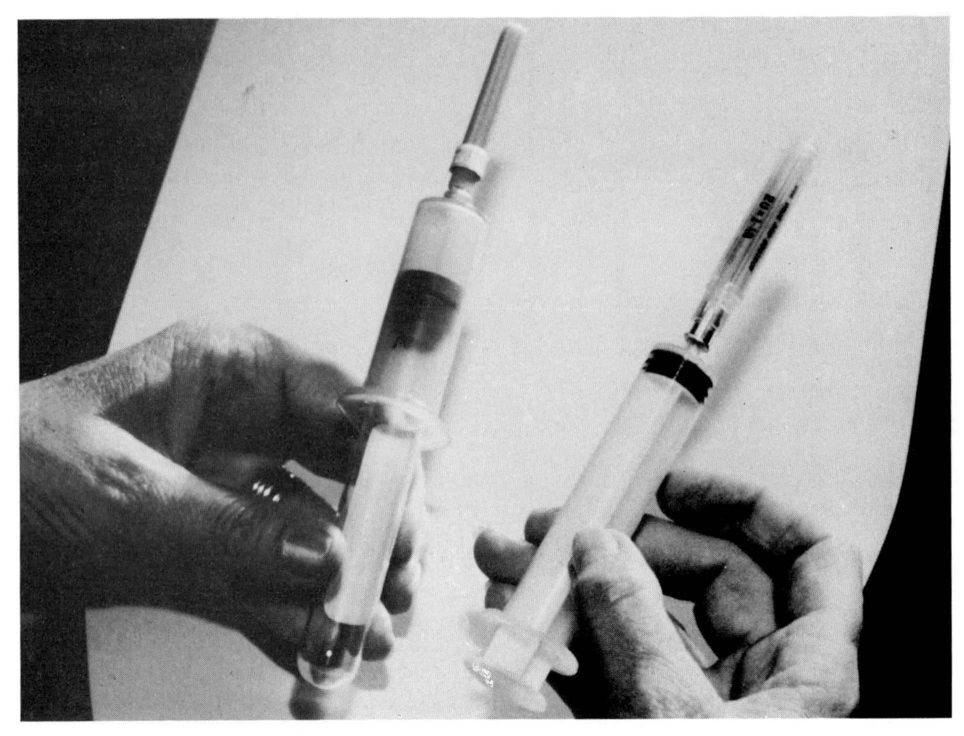

Fig. 1-5. *Evacuated system* (left) *and syringe system* (right) *for drawing blood.*

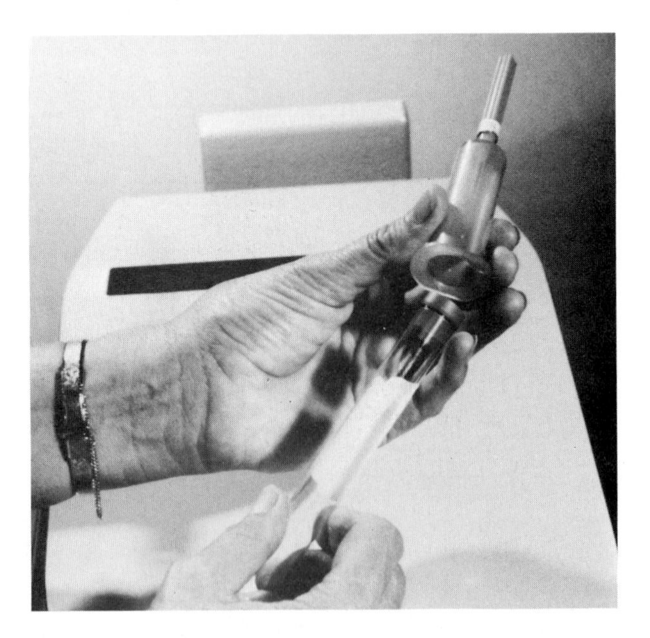

Fig. 1-6. *Three elements of evacuated system: sterile blood collection needle, holder, and evacuated tube.*

vein into the evacuated tube (Vacutainer). Evacuated tubes are also more convenient to use, are less expensive, and prevent blood from spilling when tubes are being switched. The system comprises three basic elements: a sterile blood-collection needle, a holder used to secure both the needle and the evacuated tube, and an evacuated tube that contains premeasured vacuum and premeasured additive (Fig. 1-6). Special needles are designed for the evacuated tube (Fig. 1-7); the part of the needle that screws onto the holder is called the "hub" (Fig. 1-8).

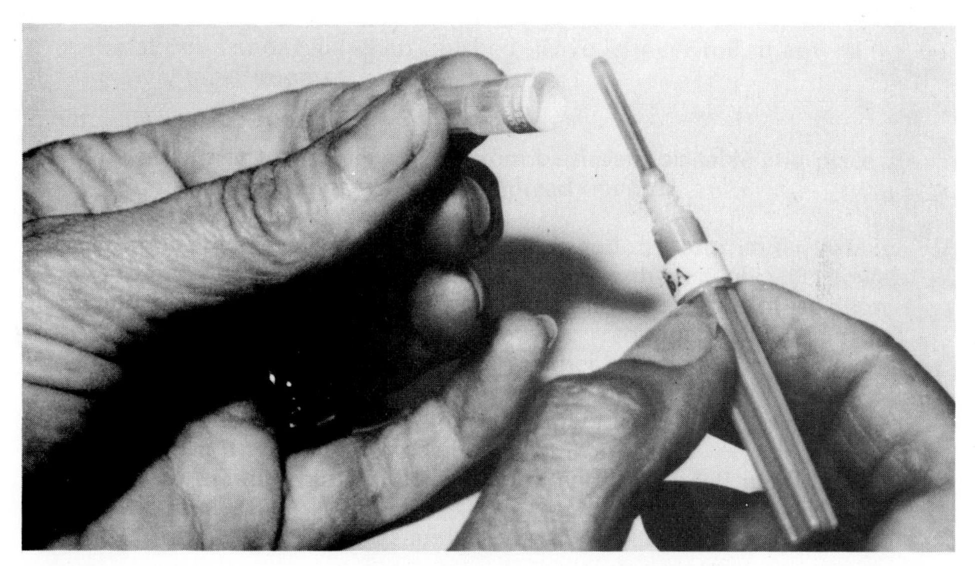

Fig. 1-7. *Portion of needle that pierces the evacuated tube stopper.*

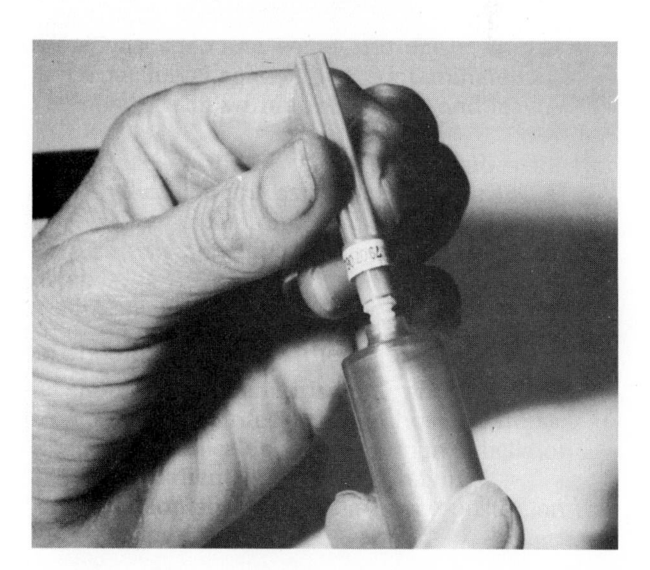

Fig. 1-8. *Threaded "hub" screws into the holder.*

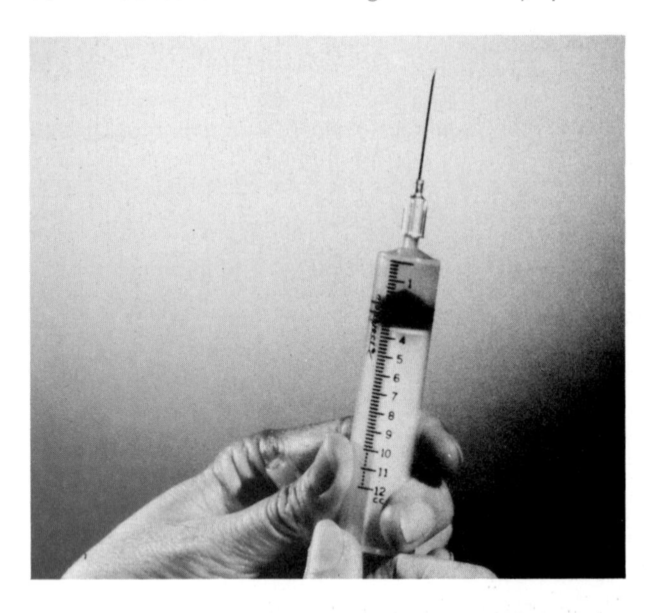

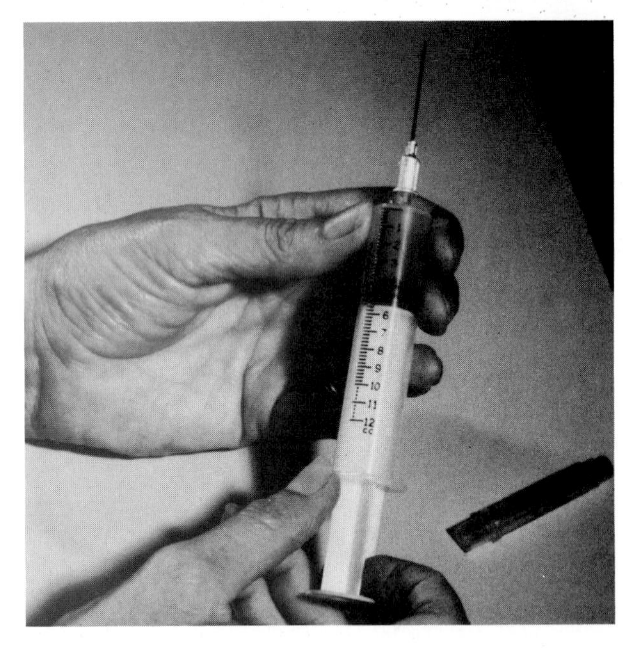

Fig. 1-9. *Needle attached to syringe* (top). *Plunger should be pulled back* (bottom) *to ensure that it is free.*

Fig. 1-10. *Color-coded stopper of evacuated tube makes checking for appropriate tube rather simple.*

Plastic or Glass Syringe

The syringe system is essentially the same as the evacuated system but does not have any vacuum; thus the blood must be drawn by pulling the plunger back from the syringe gently. Needles designed to fit onto a traditional syringe system have a different hub. After the needle has been attached to the syringe, the plunger should be pulled back to ensure that it moves freely (Fig. 1-9) and that the needle is not plugged. In general, a syringe is used only when drawing a specimen from persons with fragile, thready, or "rolly" vein walls.

CHECKING THE PAPERWORK AND TUBES

The identification number on the accessioning label must be compared with that on the blood request form. If the two numbers do not match, a further check is needed to determine which one is incorrect and to make the necessary correction. The tubes should be checked to see that the appropriate kinds and sizes have been selected, the color-coded stoppers of the evacuated tubes make this quite easy (Fig. 1-10).

Color Coding of Evacuated Tubes

Evacuated tubes come with color-coded stoppers[4] for ease of identification. Tubes for routine use can be identified by the following colors.

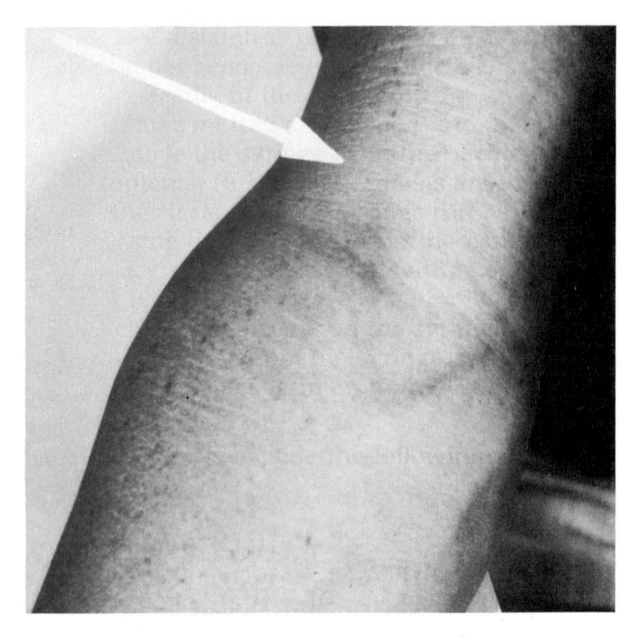

Fig. 1-11. *Median cubital vein* (arrow).

Stopper	*Additive*
Red	No additive
Green	Heparin
Light blue	Citrate
Lavender	EDTA
Gray	Fluoride or oxalate

Specialty tubes for trace metal analysis, serology studies, and fast clotting times are also stoppered in various colors for easy identification.

SELECTING VEIN SITE

For most venipuncture procedures on adults, veins located in the arm are used. The median cubital vein (Fig. 1-11) is the one used most often because it is large, close to the skin, and the least painful for the patient. If the venipuncture of this vein is unsuccessful, one of the cephalic or basilic veins (Fig. 1-12) may be used. The blood, however, usually flows more slowly from these veins; moreover, these veins tend to bruise and to roll more easily.

Before attempting to draw blood from ankle or foot vein sites, a nurse or physician should be consulted. These sites cannot be used in a patient with diabetes or a cardiovascular or circulatory problem.

In many hospitals, special identification bands indicate the need for restricted use of veins when drawing blood, such as expected intravenous therapy or insertion of a cannula.

Factors in Site Selection

Extensive Scarring. Healed burn areas should be avoided.

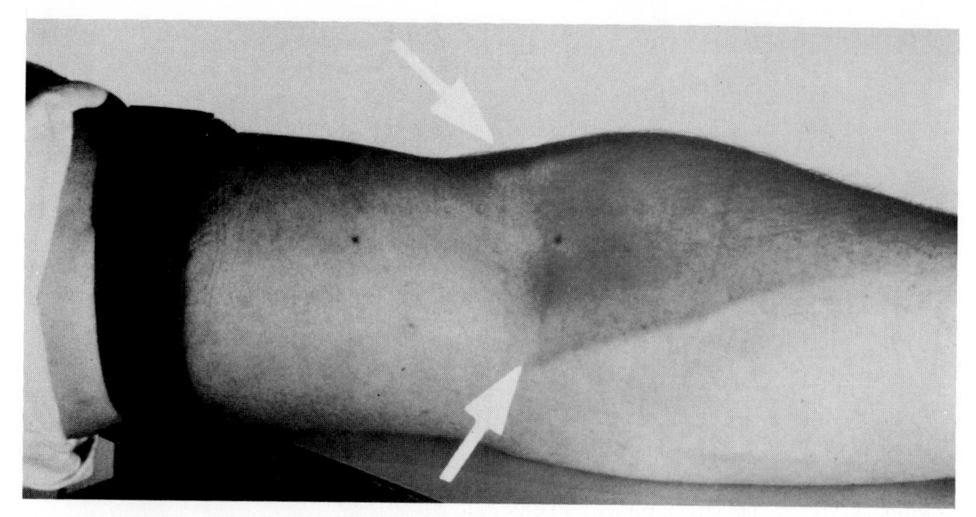

Fig. 1-12. *Cephalic vein* (top arrow) *and basilic vein* (bottom arrow).

Mastectomy. Because of lymphostasis, specimens taken from the side on which a mastectomy has been performed may not be truly representative specimens.

Hematoma. Specimens collected from a hematoma area may cause erroneous test results. If another vein site is not available, the specimen should be collected distal to the hematoma.

Intravenous Therapy. Specimens should be collected from the opposite arm; if not possible, the procedure under special situations in this chapter should be followed.

Problems in blood collection are encountered with patients who have *"difficult veins."* The following types of patients may have poor veins, making it difficult to do a successful venipuncture.

- Oncology patients, especially those receiving intravenous chemotherapy
- Leukemia patients who have had frequent blood tests
- Patients with constant intravenous therapy
- Extremely obese patients
- Babies and children
- Cardiac patients

Several techniques are useful when encountering a patient with difficult veins.

1. Look for a blood drawing site: complete forearm; back of the arm; wrists and hands (Fig. 1-13); and ankles and feet.
2. Feel for a vein using the tip of the finger because it is more sensitive. Learn to trust the sense of touch. Think of four things when feeling for a vein: bounce; direction of vein; size of needle; and depth.

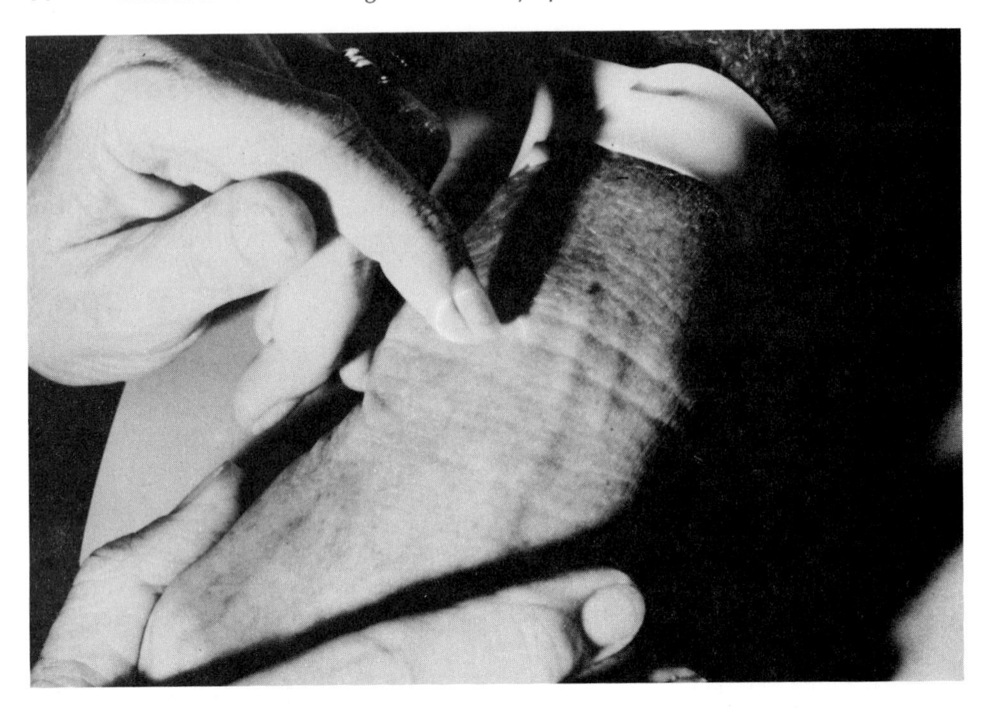

Fig. 1-13. *Looking for vein on wrist or hand.*

The patient should form a fist to make his veins more prominent and easier to enter. Vigorous hand exercise "pumping" should be avoided because it may affect some test values.

The phlebotomist's index finger palpates and traces the path of the vein several times (Fig. 1-14). Arteries, unlike veins, pulsate, are more elastic, and have a thick wall. Thrombosed veins lack resilience, feel cordlike, and roll easily. Feel firmly. Do not tap or rub finger lightly over skin because you will feel only small surface veins.

3. Choose the vein that feels fullest. Look at both arms. Always feel for the median cubital vein first; it is usually bigger, anchored better, and bruises less. The cephalic vein (depending on size) is second choice over the basilic vein because it does not roll and bruise as easily, although the blood usually flows more slowly.

The bend of the elbow is the best place to make a puncture; when this is not possible, other sites include the flexor surface of the forearms, wrist area above the thumb, volar area of the wrist, knuckle of the thumb or index finger, back of the hand, and back of the lower arm.

If unable to find a vein site immediately, the phlebotomist may attempt several procedures.

- Try the other arm unless otherwise instructed.
- Ask the patient to make a fist, which usually causes veins to become

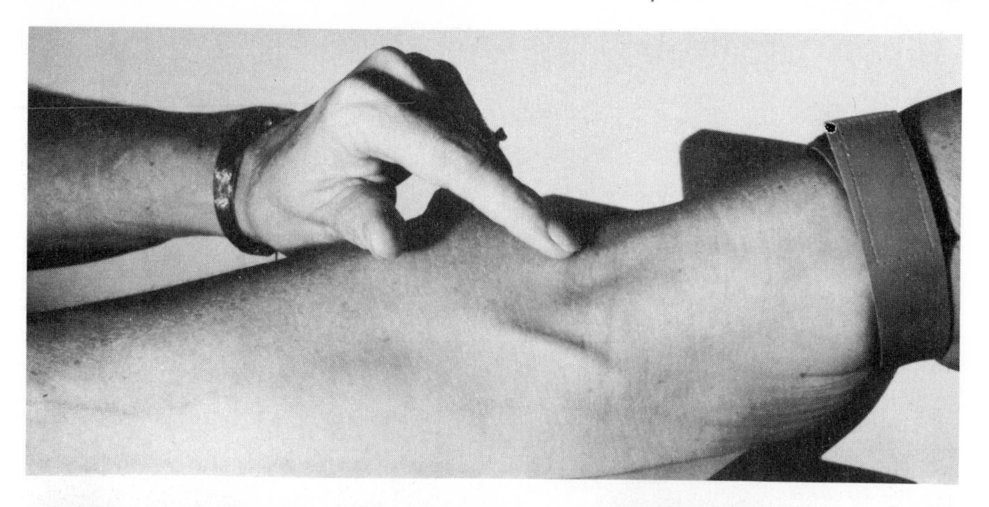

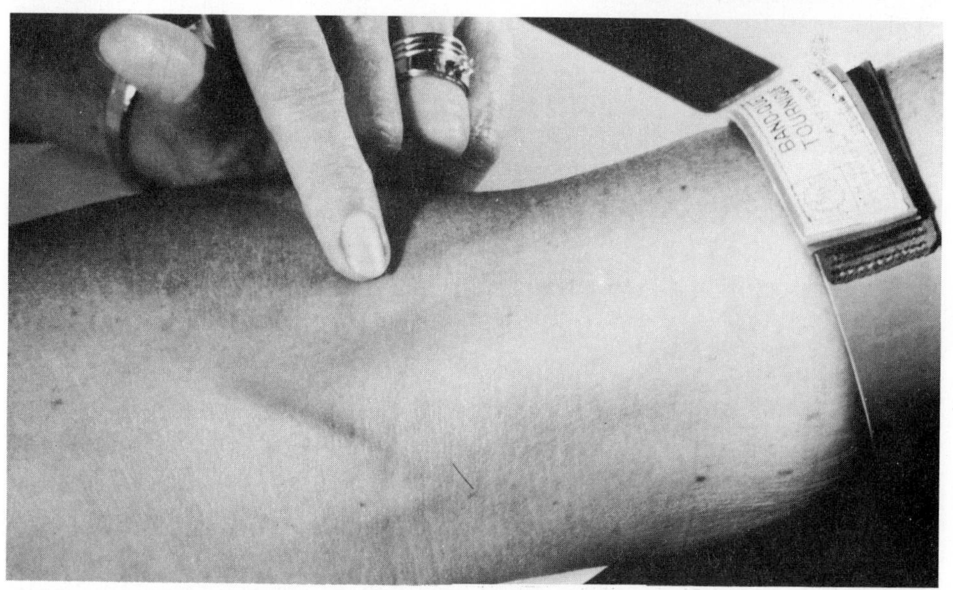

Fig. 1-14. *Index finger tracing vein path* (top, bottom).

more prominent. Sometimes it is easier for the patient to make a fist if he has a gauze roll or syringe container to grasp (Fig. 1-15).

- Apply a tourniquet briefly.
- Massage the arm from wrist to elbow (Fig. 1-16).
- Tap sharply at the vein site with your index finger a few times (Fig. 1-17).
- Apply heat to the vein site (Fig. 1-18).
- Lower the arm over the bedside or venipuncture chair (Fig. 1-19).

(*Text continued on p. 19*)

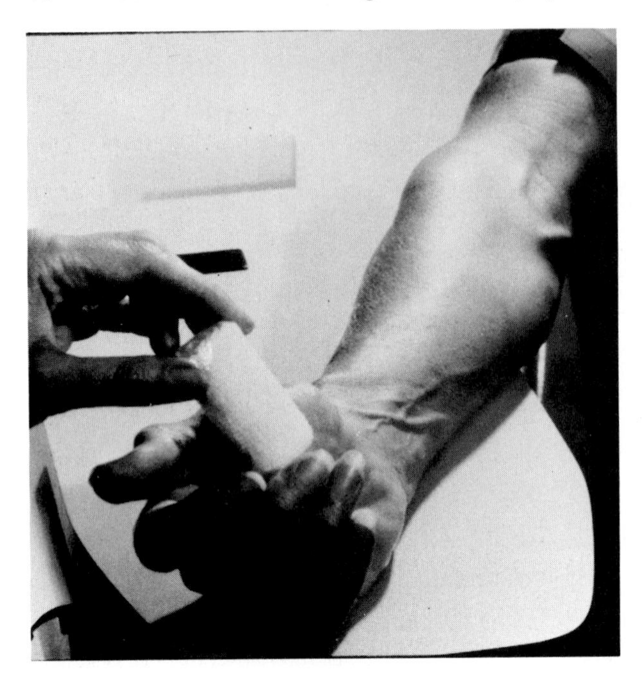

Fig. 1-15. *Patient makes a fist with gauze roll.*

Fig. 1-16. *Massage of arm to stimulate a vein.*

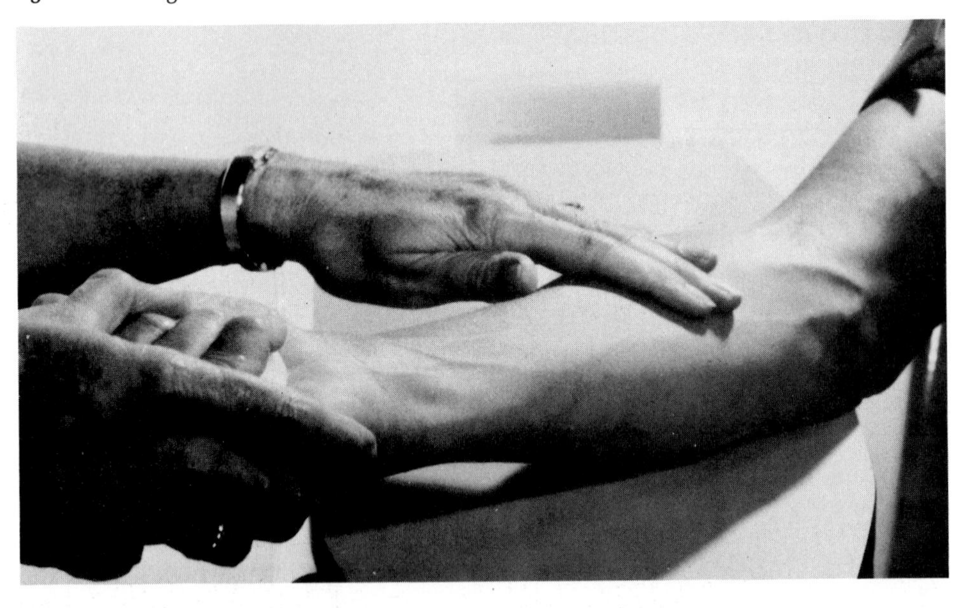

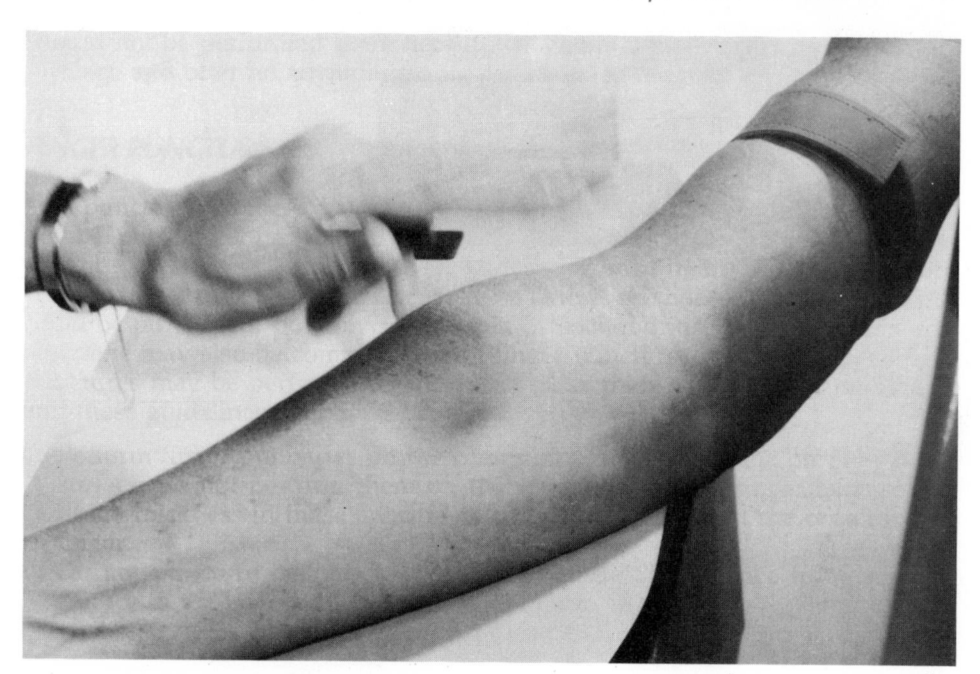

Fig. 1-17. *Tapping sharply at vein site.*

Fig. 1-18. *Heat is applied to vein site.*

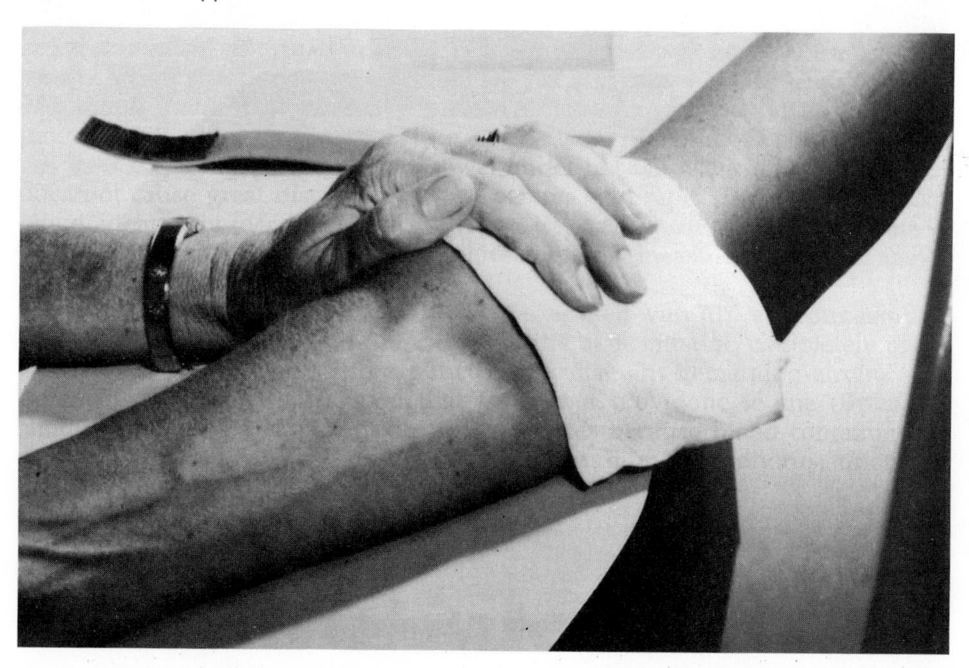

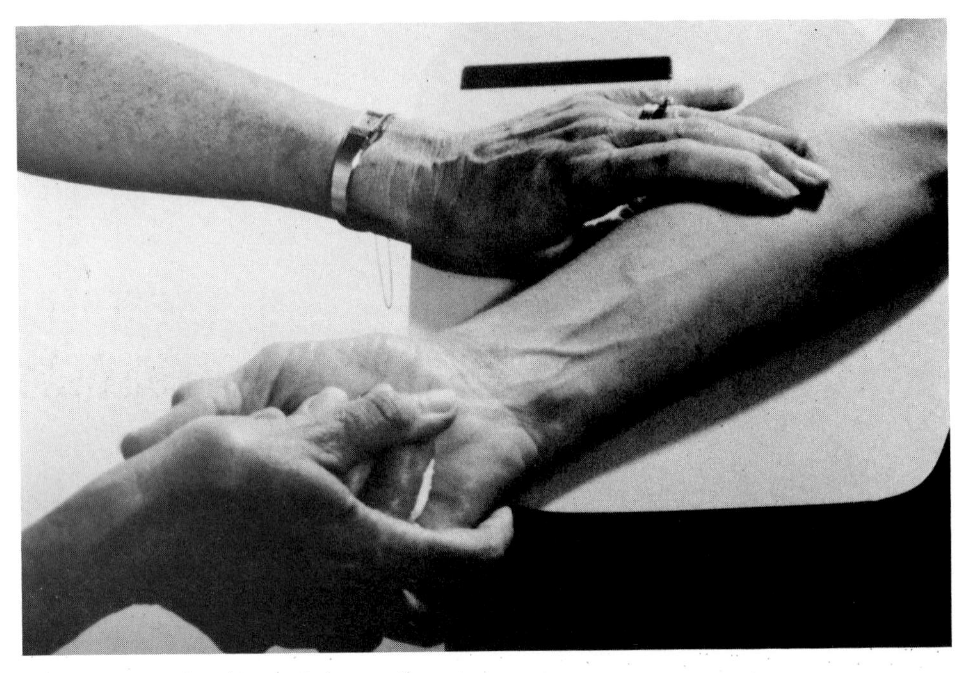

Fig. 1-19. *Lowering the arm enhances the veins.*

Table 1-1. *Basic Steps for Drawing a Blood Specimen*[3]

1. Obtain accession order
2. Ensure identification of patient
3. Check patient's diet restrictions
4. Reassure patient
5. Position patient properly
6. Assemble supplies
7. Verify paperwork and tube identification
8. Select vein site
9. Apply tourniquet
10. Cleanse venipuncture site
11. Perform venipuncture
12. Release tourniquet
13. Remove needle
14. Bandage the arm
15. Fill tubes (when syringe and needle are used)
16. Dispose of puncture unit
17. Chill specimen, if necessary
18. Eliminate diet restrictions for hospitalized patient
19. Time-stamp the paperwork
20. Send tubes to proper laboratories

APPLYING THE TOURNIQUET

A tourniquet will increase venous filling, which makes the veins more prominent and easier to enter. Two types of tourniquets are available: the soft rubber bandage, 1.25 cm (½ in) side and 45 cm (18 in) long, and the Velcro-type band (Fig. 1-20). The rubber bandage should be wrapped around the patient's arm rather tightly and the end tucked under the last round (Fig. 1-21); the Velcro band should be wrapped firmly around the patient's arm and the ends secured (Fig. 1-22). The rubber bandage, with its tendency to pinch the skin of obese patients, especially, should be wrapped carefully starting about 7.5 to 10 cm (3–4 in) above the site (Fig. 1-23). As the tourniquet increases pressure, the vein becomes enlarged, which will aid in locating and entering the vein. For valid test results, the tourniquet should never be left on the arm for more than 2

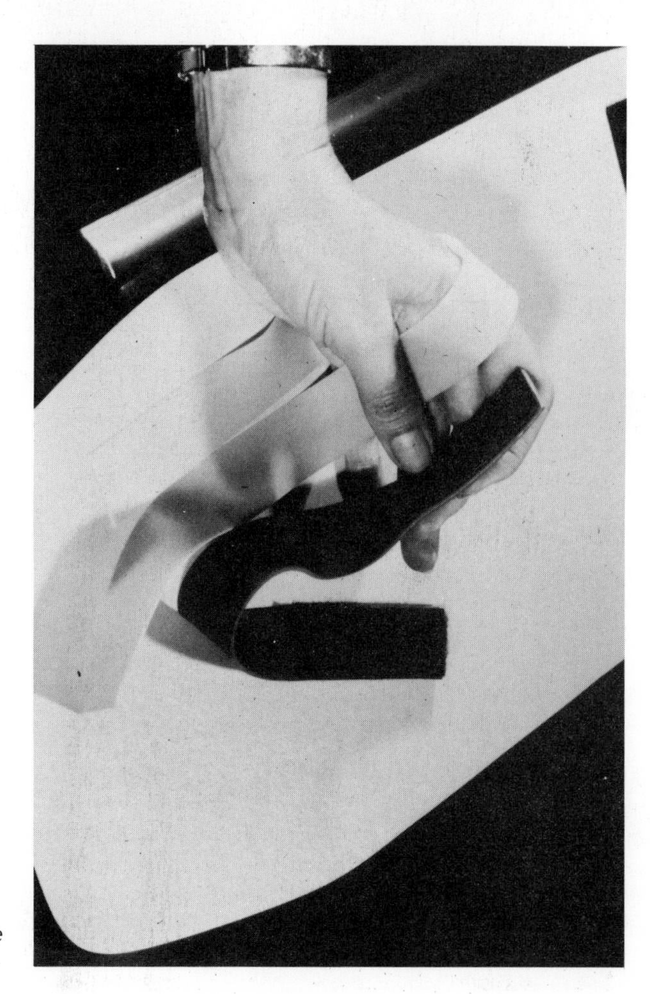

Fig. 1-20. *Two types of tourniquets: soft rubber bandage* (top) *and Velcro-type band* (bottom).

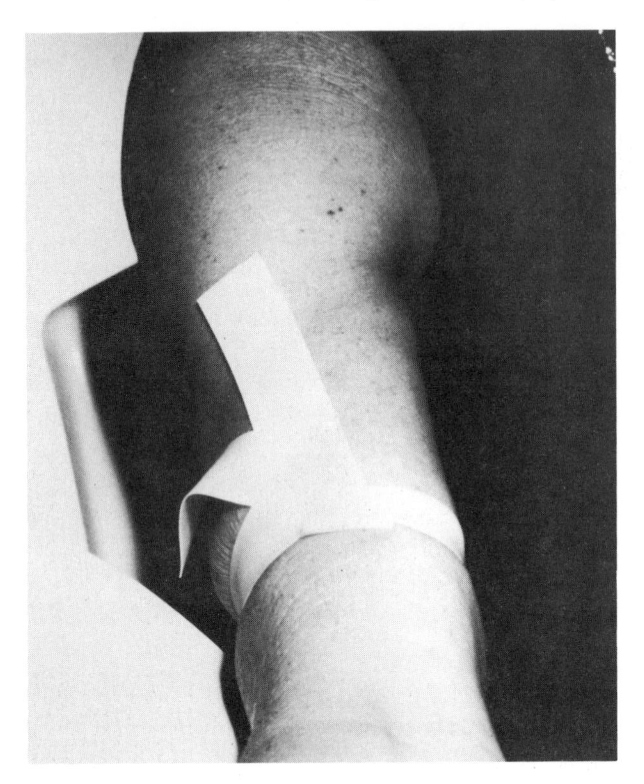

Fig. 1-21. *Rubber tourniquet wrapped around patient's arm.*

minutes; because a tourniquet prevents the blood from flowing freely, the balance of fluids and blood elements may be disrupted if the tourniquet is left on too long.[5] When a patient cannot make a fist, a second tourniquet is applied below the proposed puncture site and released, in most cases, as soon as the blood is flowing. The Velcro tourniquet is good because it can be loosened easily with one hand.

CLEANSING THE AREA

Once the vein to be used has been located, the phlebotomist must cleanse the area thoroughly to prevent any contamination. He first soaks a gauze pad in 70% isopropyl solution and then moves the pad in a circular motion from the center of the vein site outward (Fig. 1-24). The area is allowed to dry to prevent possible hemolysis of the blood specimen, which would affect the test results, and to avoid a burning sensation to the patient when the venipuncture is performed. If the skin is touched after it has been cleansed, the procedure must be repeated.

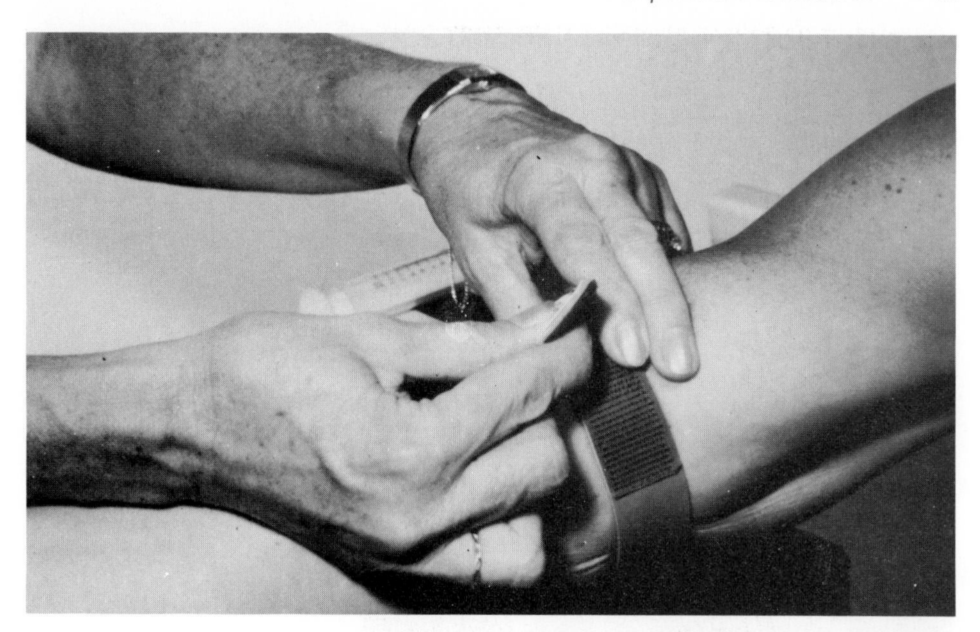

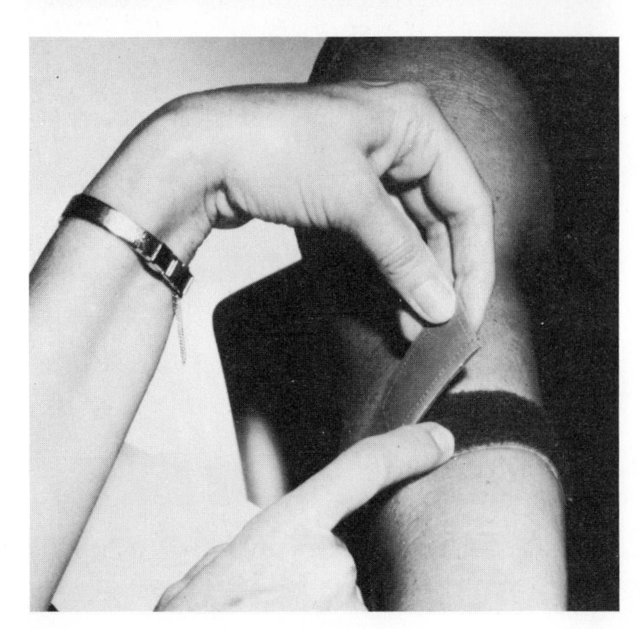

Fig. 1-22. *Velcro tourniquet wrapped around patient's arm* (top) *and the ends secured* (bottom).

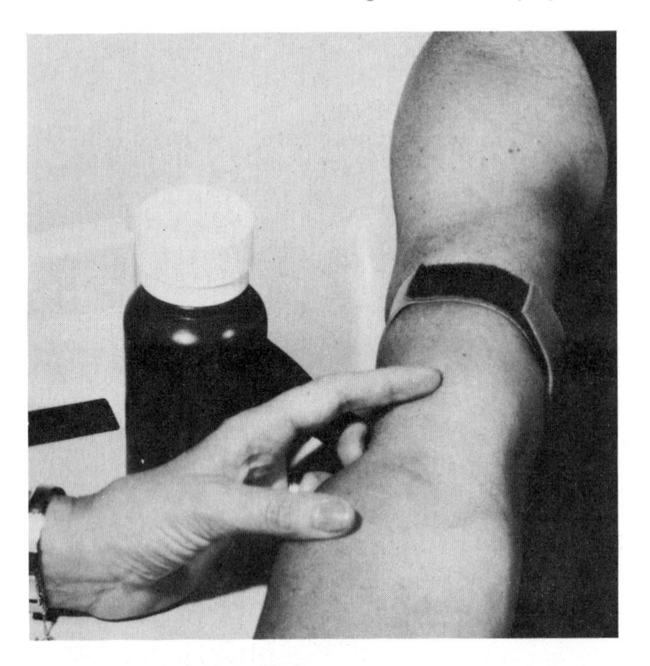

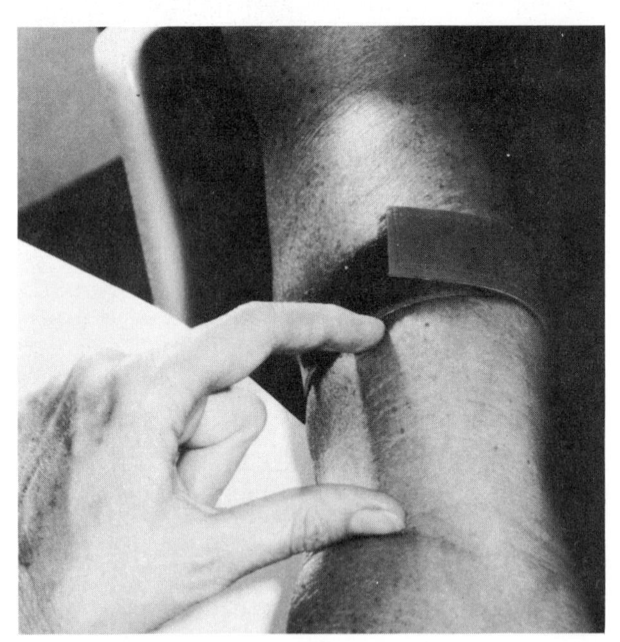

Fig. 1-23. *Application of tourniquet about 7.5 to 10 cm (3–4 in) above the site* (top, bottom).

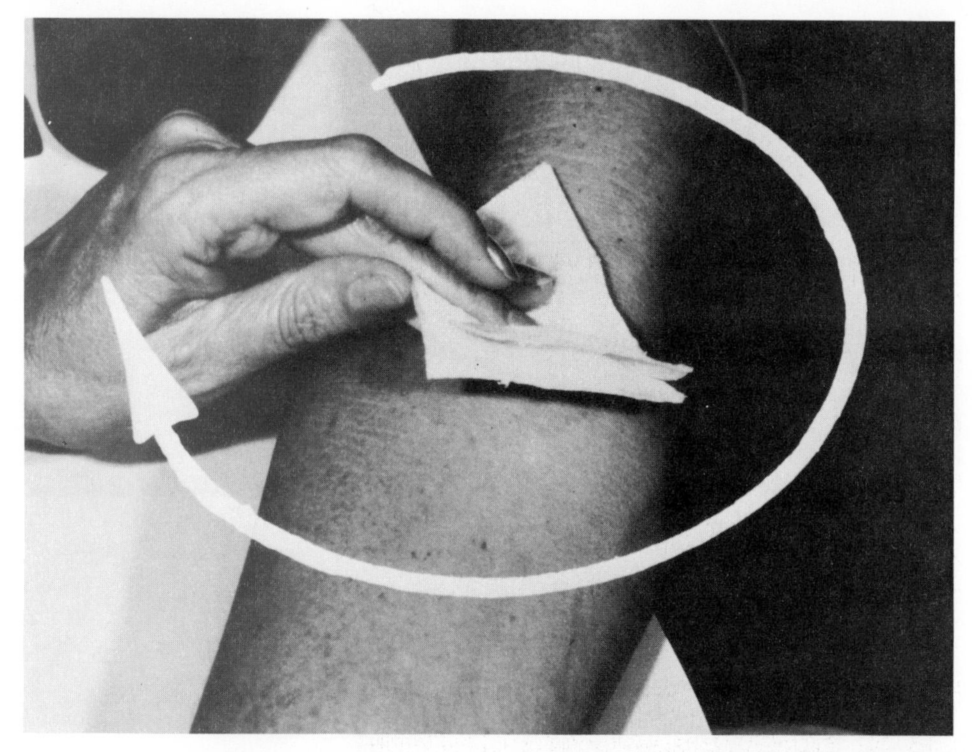

Fig. 1-24. *Cleansing of area with alcohol-soaked gauze pad. Pad should be moved in a circular motion from center of vein site outward.*

INSPECTING THE NEEDLE, SYRINGE, OR EVACUATED TUBE

The appropriate needle is attached to the syringe or evacuated tube holder and threaded into the holder until it is secure, using the needle sheath as a wrench. The cover of the needle must not be removed until the phlebotomist is ready to draw blood, thus avoiding needle contamination. When ready for use, the needle should be removed from the cover and examined, especially the tip to determine if it is free of hooks at the end of the point and if its opening is clear of small particles that would obstruct the blood flow.

The blood collector is then inserted into the holder but should not be pushed onto the needle because the vacuum may be lost prematurely.

PERFORMING THE VENIPUNCTURE

Evacuated Tubes

The phlebotomist should grasp the patient's arm firmly to facilitate the venipuncture procedure, using the thumb to draw the skin taut (Fig. 1-25) and thus anchoring the vein. The vein is entered with the bevel of the needle upwards (Fig. 1-26). Initially, some resistance is encountered, but once the point of the needle passes through the vein wall a release is felt. One hand should hold the

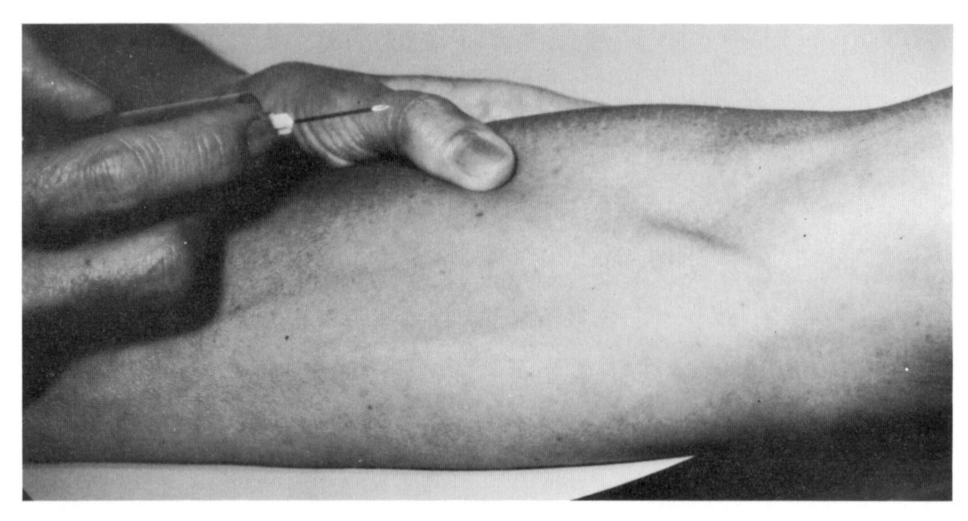

Fig. 1-25. *Grasping of patient's arm, using thumb to draw skin taut.*

Fig. 1-26. *Entering vein. Bevel of needle should be pointed upwards.*

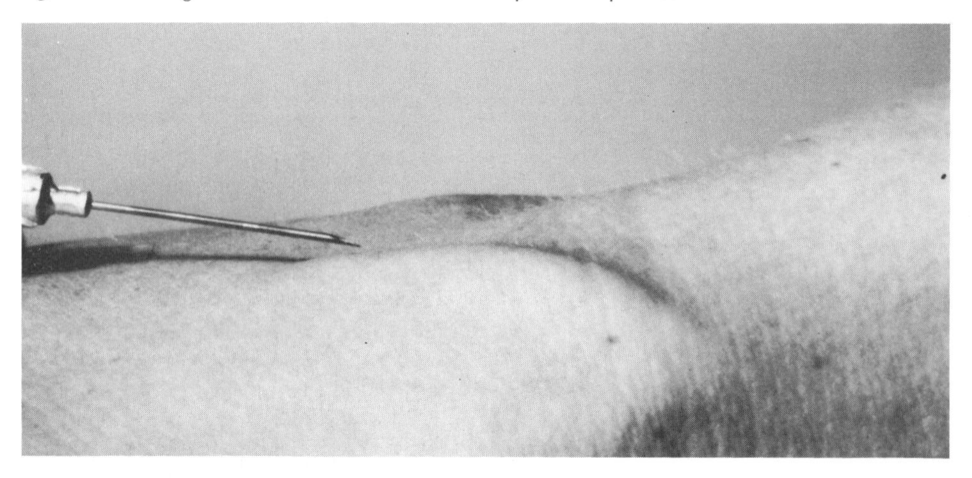

evacuated tube while the other depresses the tube to the end of the holder (Fig. 1-27). The butt end of the needle will puncture the stopper, activating the vacuum action to draw blood (Fig. 1-28).

The tube should be filled until the vacuum is exhausted and the blood flow ceases, thus ensuring a correct ratio of anticoagulant to blood. It is normal for the tube not to be filled completely (Fig. 1-29); after the tube has been filled completely, it should be removed from the holder (Fig. 1-30). The shut-off valve recovers the point, stopping the blood flow until the next tube is inserted. After drawing each tube, an additive should be mixed immediately by inverting the tube three or four times. Gentle inversion will avoid hemolysis.

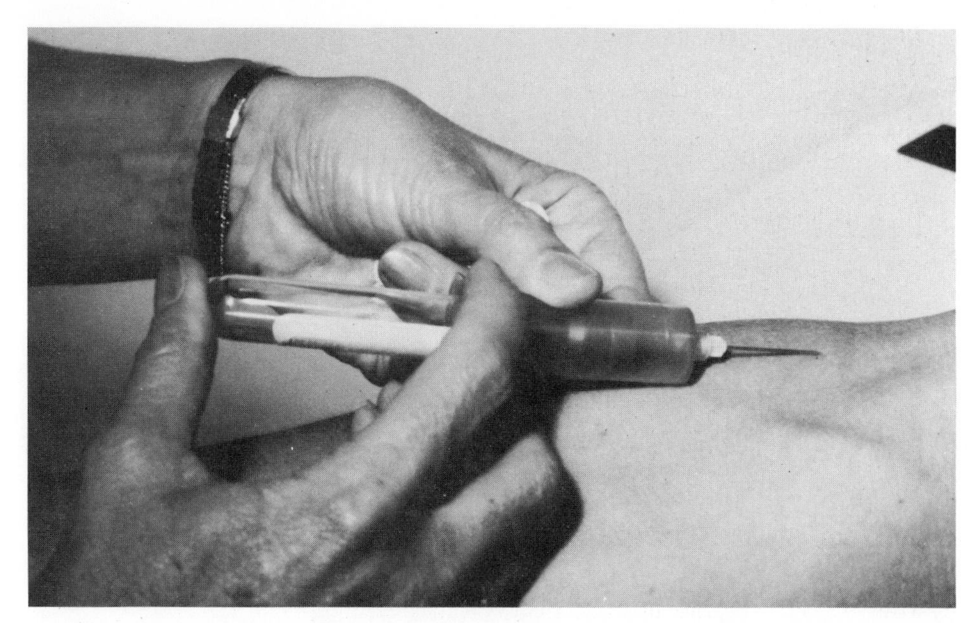

Fig. 1-27. *One hand holds evacuated tube while other hand depresses tube to end of holder.*

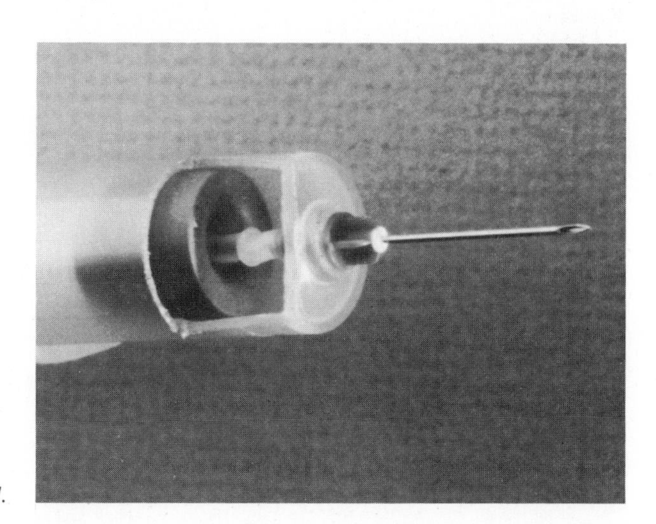

Fig. 1-28. *Butt end of needle punctures stopper, activating vacuum action to draw blood.*

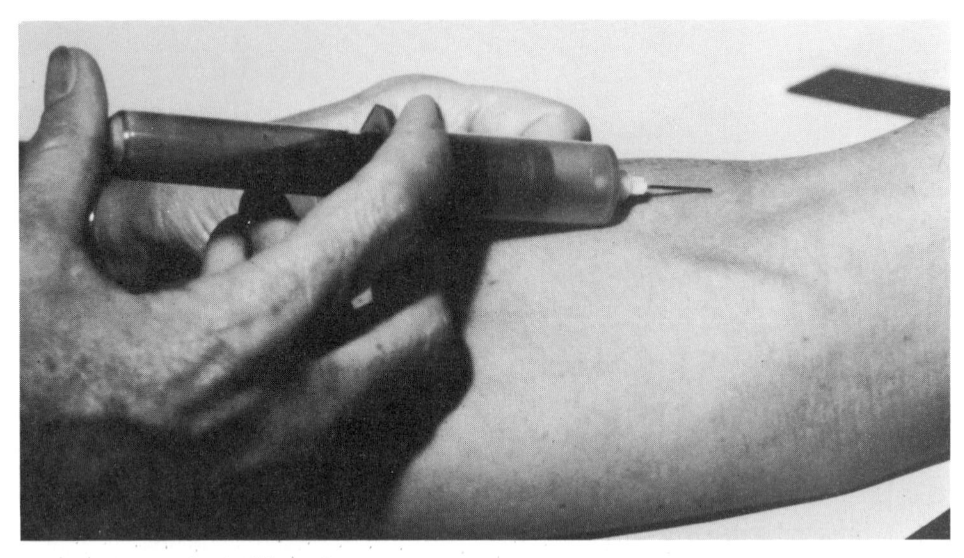

Fig. 1-29. *Incompletely filled tube.*

Fig. 1-30. *After tube has been filled, it should be removed from holder before needle is withdrawn.*

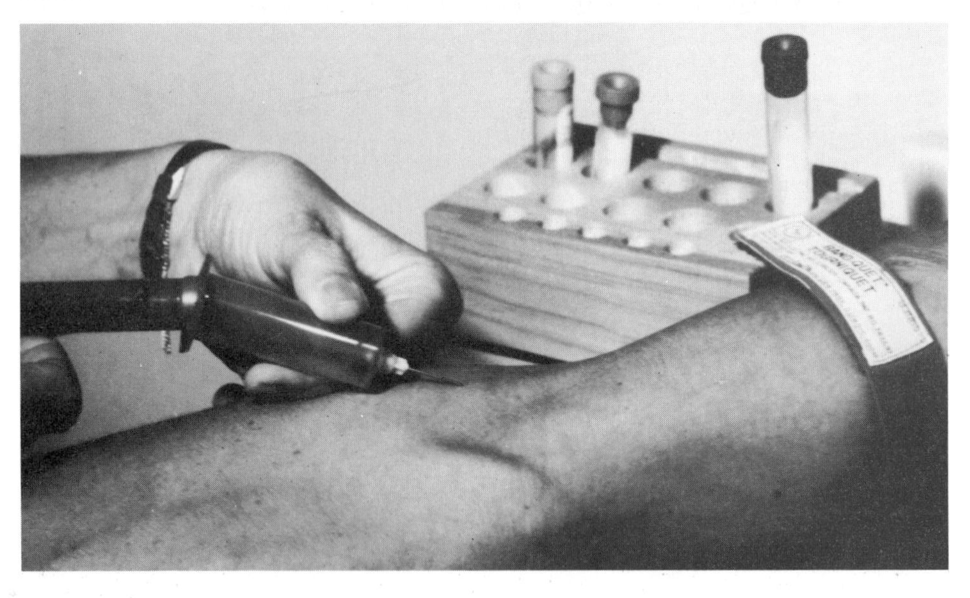

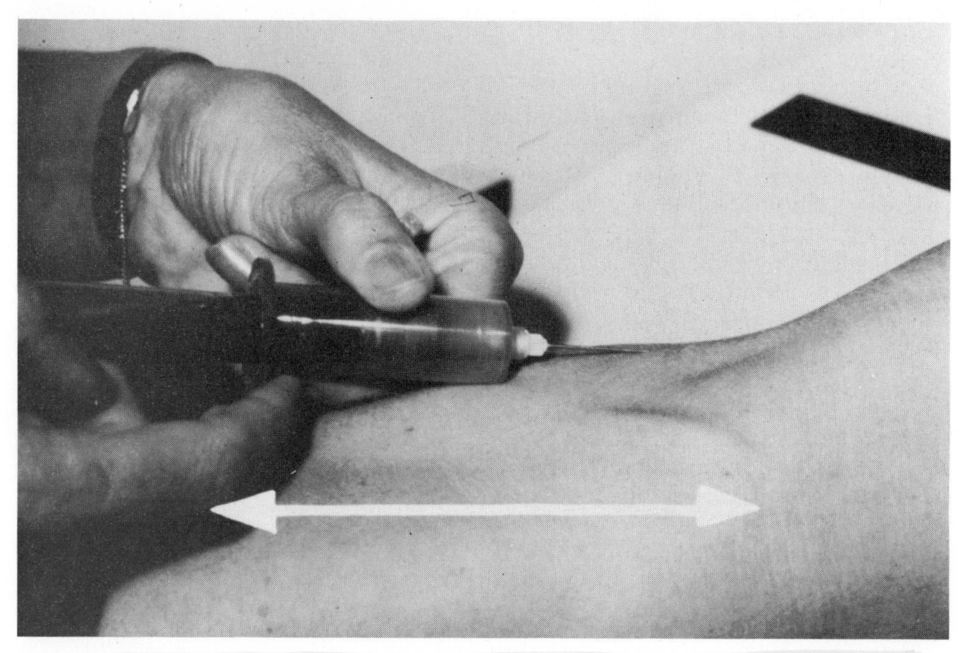

Fig. 1-31. *Backward and forward movement of needle sometimes increases the blood flow.*

Occasionally, a faulty tube will have no vacuum. If a tube is not filling and the needle is inside the vein, another tube should be used. If a tube starts to fill but then stops, the needle should be moved slightly forward or backward (Fig. 1-31); usually this adjustment will increase the flow. The needle is then rotated half a turn, and the tourniquet, which may have been applied too tightly, is loosened. Probing is not recommended because it is painful to the patient. If none of these procedures is helpful, the needle should be removed and an alternate site used.

Syringe

A syringe and needle almost always are used to collect blood from patients with difficult veins. If a puncture has been made and the blood is not flowing, the venipuncturist should determine if he is pulling too hard on the plunger and collapsing the vein. The needle should be drawn back while the plunger is being pulled slightly. Keeping the needle in the patient's arm, the venipuncturist must make sure that the bevel is covered by the skin. With the syringe in his right hand, he uses the index finger of his left hand to feel for the vein. After the vein is relocated, the finger is kept gently on the vein and the needle guided to that point. The venipuncturist then pulls gently on the plunger. As soon as the blood starts to flow into the syringe, the needle should not be moved.

Patients who have had chemotherapy may have scarred veins; it may take longer to find a usable vein in these patients. Edema may complicate blood collection, and punctures often have tissue obstructing the needle so that it is necessary to restick. With cardiac patients, especially children who are cyanotic (lips, fingernails), the needle size is important because the blood is very thick and will not pass through a small needle readily. In obese persons with double creases at the elbow, a vein may be found on the inner (basilic) side of the lower crease. Often the veins in obese persons are not as deep as they feel.

When all possibilities have been exhausted and the procedure still is unsuccessful, a coworker's aid should be enlisted.

RELEASING THE TOURNIQUET

After blood has been drawn, the patient opens his hand and the tourniquet is released (Fig. 1-32), allowing blood circulation to return to normal and reducing the amount of bleeding at the venipuncture site. The venipuncturist folds a gauze pad in fourths and holds it over the needle, which is then gently and carefully removed. The gauze pad is held firmly over the venipuncture site (Fig. 1-33) and the arm wrapped firmly three or four times with a gauze bandage over the pad (Fig. 1-34). The patient may remove the bandage in 15 minutes. Bandaging helps to prevent bruising and is a comfort to the patient.

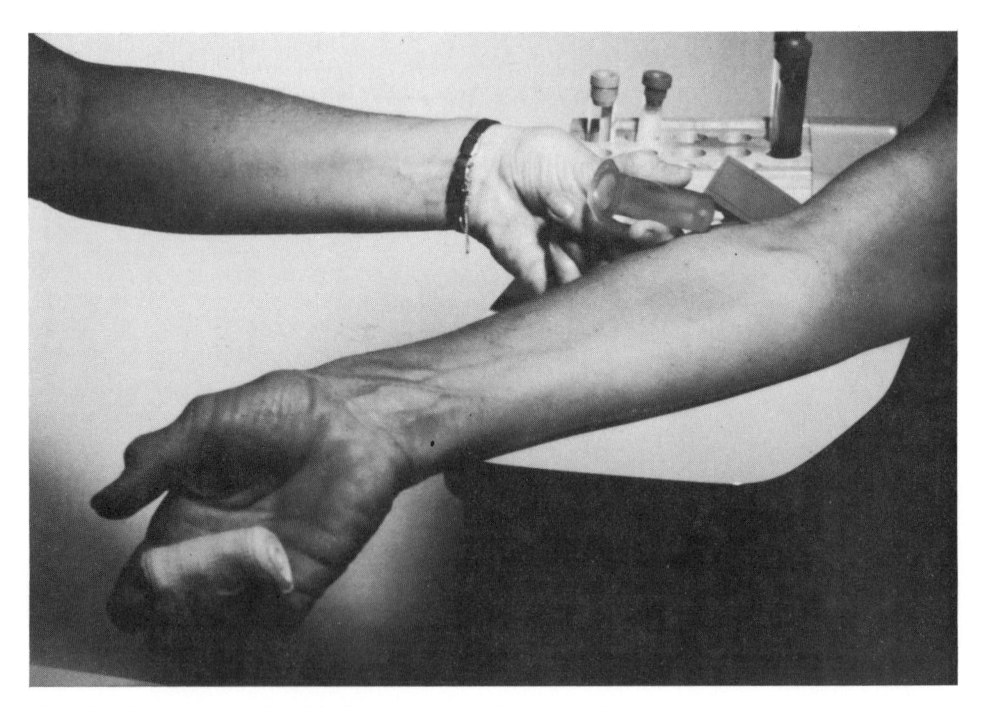

Fig. 1-32. *Patient opens hand before tourniquet is removed.*

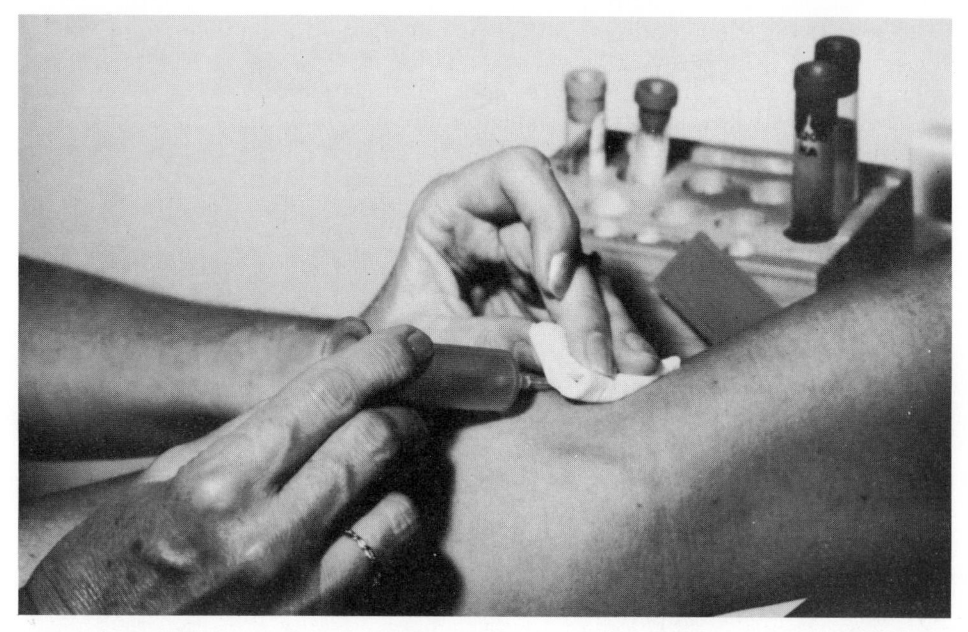

Fig. 1-33. *Gauze pad placed and held firmly over vein puncture site.*

If a patient continues to bleed, pressure is applied to the site with a gauze pad until the bleeding stops. Patients on anticoagulant therapy will often need more time to stop bleeding.

The needle is removed from the syringe or evacuated tube and destroyed using a needle cutter box (Fig. 1-35).

FILLING THE APPROPRIATE TUBES OF SYRINGE-DRAWN SPECIMENS

If evacuated tubes are filled from a syringe, stoppers should not be removed. The diaphragm of the rubber stopper on the appropriate tube must be punctured and the correct amount of blood allowed to flow slowly into the tube. Blood should never be forced into a tube. If the tube does not fill, the plunger of the syringe may be pushed gently—an extremely important technique.

CHILLING THE SPECIMEN (FOR CERTAIN SPECIMENS ONLY)

Some tests require that the blood be cooled immediately after the venipuncture to slow down metabolic processes that may cause altered chemical values. Some tests that require chilling of the specimen are serum gastrin, ketone, lactic acid, activated partial thromboplastin time, renin activity, and vitamin C.

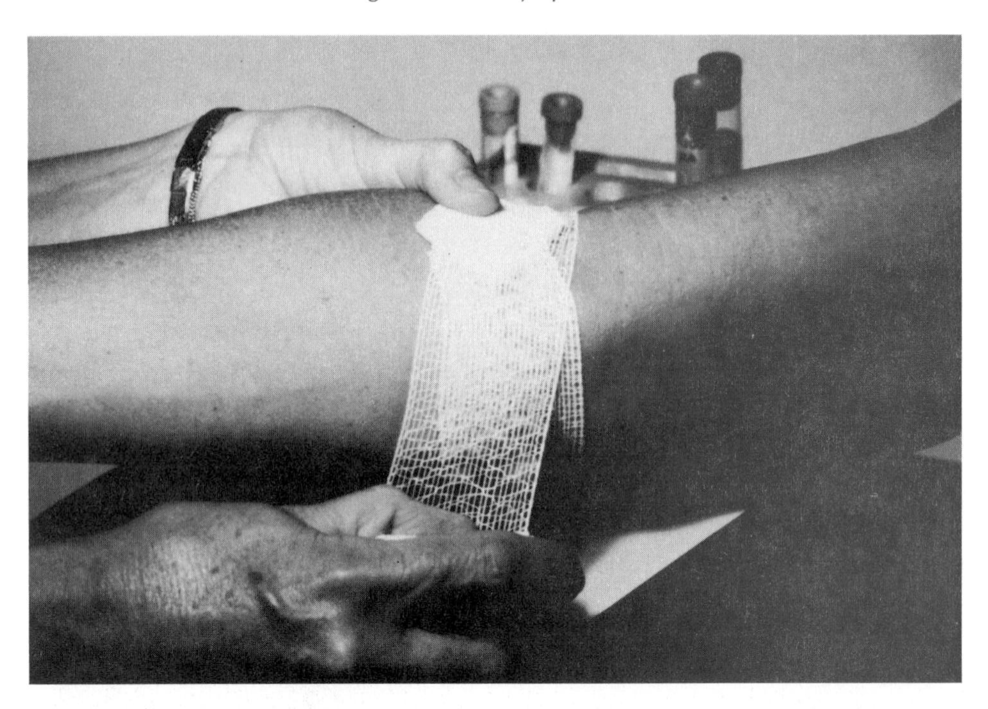

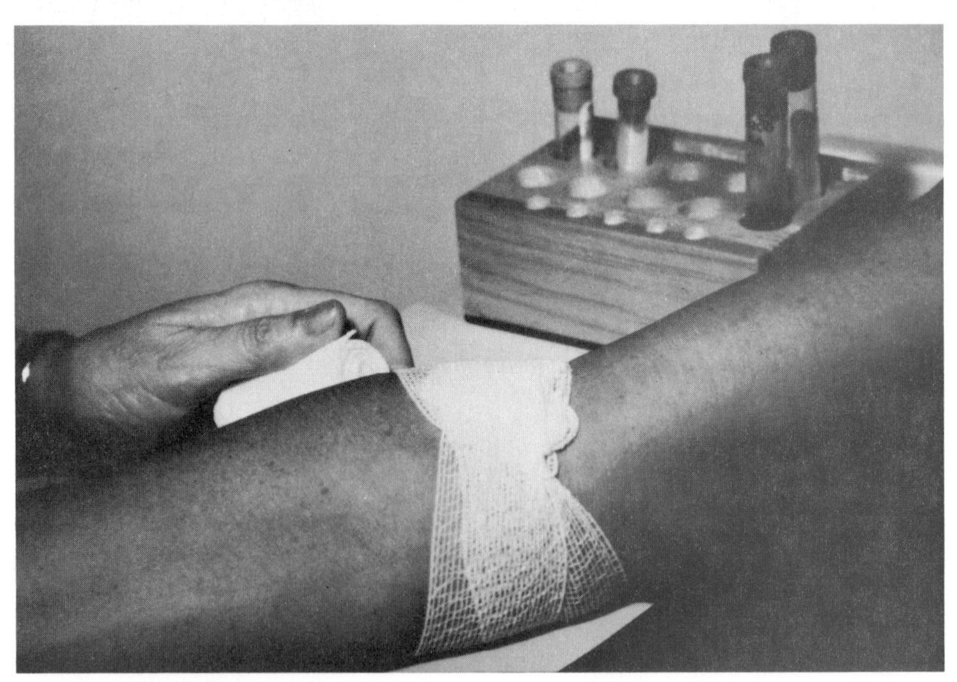

Fig. 1-34. *Gauze bandage should be wrapped around the arm three or four times over the gauze pad* (top, bottom).

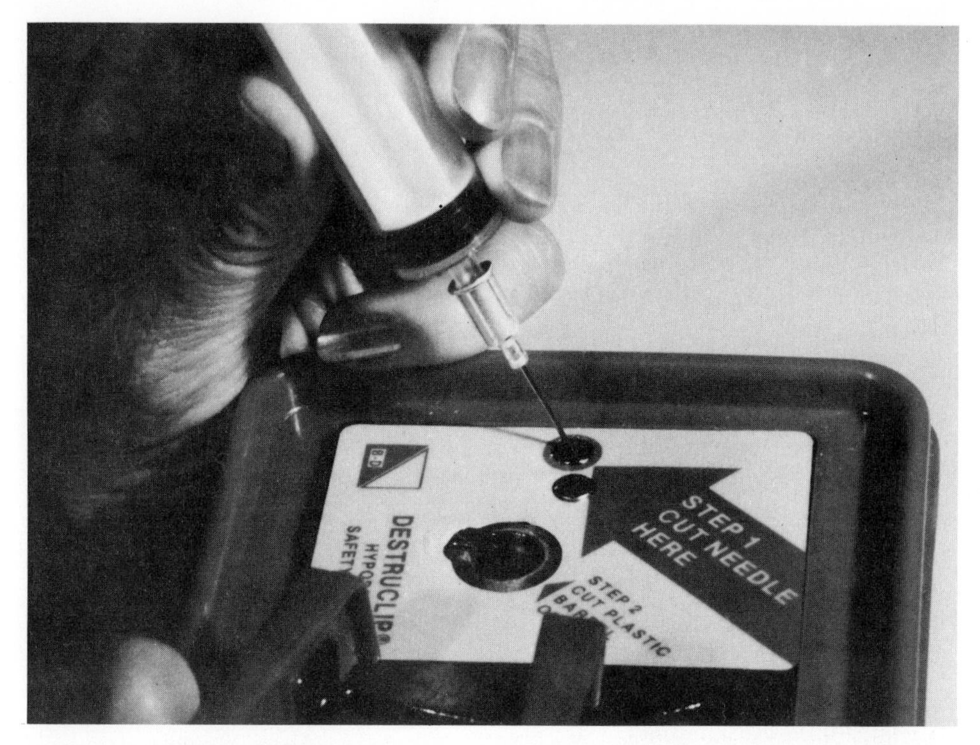

Fig. 1-35. *Destruction of needle.*

MARKING THE PATIENT'S NAME OFF "HOLD BREAKFAST" LIST

In the hospital, the patient's name should be removed from the "hold breakfast" list once blood has been drawn. This will notify hospital personnel that the patient may have breakfast served. In the outpatient area, if the patient inquires about having breakfast, he is told to check his card for other appointments that may require him not to eat.

TIME-STAMPING OF TEST FORMS AND ENVELOPES

In the hospital, once the specimen has been drawn, the venipuncturist should time-stamp each test form and the envelope in which the forms are placed. The time-stamp machine should stamp the month, day, year, and time on each form. When computers are used, the same information should be entered for the patient to provide a permanent record for the physician who needs to know exactly when each specimen was drawn to correlate the results with any change in the patient's condition. The laboratory also needs to know when the specimen was collected. The tubes should be sent to the appropriate laboratories designated to perform the tests.

ADDITIONAL CONSIDERATIONS

Drawing Order

It is important to draw sterile blood culture specimens first, then specimens that require no additives, followed by coagulation specimens, if done at the same time, and then specimens that are collected in tubes with additives. A recent paper suggests that there can be *additive* contamination of specimens from tube to tube when there is difficulty with the venipuncture.[6]

Each institution should develop its own order of draw because policies on coagulation studies and blood cultures may vary according to rules of the laboratory.

Preventing a Hematoma During Venipuncture

- Puncture only the uppermost wall of the vein.
- Remove the tourniquet before removing the needle.
- Use the major veins.
- Make sure the needle fully penetrates the uppermost wall of the vein. Partial penetration may allow blood to leak into the soft tissue that surrounds the vein by way of the needle bevel.
- Apply a small amount of pressure to the area with the gauze pad when bandaging the arm.

Hemolysis in Blood Samples

Hemolysis is the liberation of hemoglobin after red blood cells have ruptured. When hemolysis occurs, the serum or plasma has a pink to red color.

Causes of Hemolysis

Hemolytic anemia, liver disease, and transfusion reactions (incompatible blood) are among the causes of hemolysis. Another is the manner in which the blood sample has been drawn. In venipuncture, for instance, hemolysis may occur from

- using too small a needle;
- forcing the blood through the needle into the tube, which often causes frothing and, in turn, hemolysis;
- forcing blood from a syringe into a tube when the blood is starting to clot;
- shaking the tube of blood too vigorously instead of inverting it gently;
- drawing blood from a hematoma;
- pulling on the plunger of the syringe too forcefully;
- centrifuging the blood specimens before the blood has clotted completely.

In microblood specimens, hemolysis may occur from

- alcohol left on the skin, which mixes with blood to cause hemolysis;
- squeezing the heel or finger too hard, which causes the red cells to rupture.

Hemolysis of blood from newborns may be greater than that from blood obtained by venipuncture because of increased red cell fragility and high hematocrit levels.

Some of the tests affected by hemolysis include the following.

- Bilirubin
- Chloride
- Cholesterol
- Chemistry group
- Phosphorus
- Hematology group
- Haptoglobin
- Iron
- Most enzymes, especially
- Phosphatase—acid
- Phosphatase—alkaline
- Lactic dehydrogenase

SPECIAL SITUATIONS

INTRAVENOUS FLUIDS

"Intravenous solution" defines any fluid administered by injection into a vein, usually by a regulated-drip technique. The type of fluid that may be administered includes saline or glucose solution, one of many combinations of chemicals and water, or a transfusion of whole blood or one of its components. The amount of fluid may vary from milliliters to several liters.

When a patient is being administered an intravenous (i.v.) infusion (normally in an arm), blood should not be drawn from that site unless no other site is available. Blood drawn above an i.v. site will be diluted with the fluid being administered. Test results from this blood will be erroneous and thus misleading to the physician. One should look for a blood drawing site in the opposite arm.

Occasionally i.v. lines will be running in both arms, and no site can be found except in the area of i.v. administration. Satisfactory samples may be drawn *below* the i.v. site if the following procedure is followed.

1. Ask the nurse to turn off the i.v. line (nursing personnel must be aware if administration of fluid has been interrupted).
2. Wait 2 minutes. Apply tourniquet below the i.v. site. Select a vein other than the one with the i.v. line.
3. Perform veniuncture. Draw 5 ml of blood and discard.
4. Draw test sample. After withdrawal of needle, apply a firm, but not tight, bandage.
5. Ask the nurse to restart the i.v. line. (The venipuncturist must never restart the i.v. line. The drip rate is critical in many cases.)
6. Inform the laboratory and the physician about the procedure that was followed, using the report form. Example: Drawn below i.v. site; i.v. off 2 minutes; 0.9 N/saline running.

OBTAINING BLOOD SAMPLES FROM INDWELLING LINES

A *line* is a piece of tubing inserted into a vein or artery to administer fluids and medications, to monitor pressures, and to obtain blood samples for diagnostic tests. Types of lines used for obtaining blood samples are internal jugular, arterial, umbilical, and, occasionally, atrial and central venous pressure system.

After blood has been withdrawn from lines, the lines must be flushed with a heparin solution immediately to reduce the risk of thrombosis. Meticulous sterile technique must be observed to reduce the chance of bacterial contamination. Only physicians, nurses, and specially trained personnel should be responsible for performing this task.

The venipuncturist should not draw from the lines. His responsibilities should be limited to telling the nurse or monitoring technician what tests have been ordered, the volume of specimen needed, the patient's identification, the proper distribution and identification of the specimen, and the source of the sample on the test form. Thus, the drawing of specimens from lines requires the participation of two persons: One procures the sample and takes care of the indwelling line after the specimen is collected, whereas the other person distributes the specimen properly and identifies the patient and the specimen tubes correctly.

Lines must be cleared of fluid before blood specimens can be drawn for diagnostic testing. Two syringes are used for this procedure: The fluid and blood withdrawn in the first syringe is discarded, but the blood in the second syringe is used for testing. At least three times the volume of fluid that was in the lines must be drawn and discarded. Four to five times the volume should be withdrawn if coagulation tests have been ordered because even minute amounts of heparin may alter test results.[7] The source of the specimen should always be indicated on the test form so that the physician can evaluate results properly.

Meticulous sterile technique must be observed by the person collecting the specimen. He must remove and discard at least 5 ml of fluid and blood from the line before collecting the test sample. After the collection, he must flush the indwelling line with an appropriate solution. The venipuncturist should distribute the sample, label it properly, and identify the source of the sample on the test forms.

The decision to draw test specimens from lines, and which tests to draw, should always be made by the attending physician and observed by the venipuncturist. The decision usually is based on the need for frequent monitoring of certain blood values for proper treatment and the unavailability of veins in the extremities because of trauma, veins being used for monitoring, or i.v. therapy and medication.

HEPARIN LOCKS

An indwelling winged-butterfly needle can be left in a vein from 36 to 48 hours to administer medication intravenously or as a vein source to obtain a blood sample. Known as a *heparin lock,* this system is being used more widely in hospitals to "save" veins for therapeutic use and to lessen trauma to the patient. Repeated venipunctures not only are painful to the patient but also cause scarring of the lining of the vein, which may render that vein unusable. With this system, a continuous i.v. infusion is not needed to keep the vein open, thus giving the patient more comfort and mobility.

The butterfly system must be placed very carefully in the vein and the site maintained during its use because this needle is a foreign body injected directly into the patient's vein and may lead to infection. Before placing the heparin lock into the vein, the site should be shaved and cleansed well with an antiseptic solution. During the bandaged period of use, an antibiotic ointment should be applied over the needle site and monitored for signs of inflammation. If the skin becomes painful, red, or swollen, the needle should be removed and hot, moist heat applied to the area.

Prepared heparin flush or heparin diluted 1:1000 in saline is used in the line to keep it from clotting. The solution is injected through the tubing and a plug used at the end of the butterfly line to hold this solution in place. Butterfly needles specially designed for use as heparin locks are now available with a removable plug built into the system. Before the actual blood specimen is withdrawn, a waste specimen of 2 ml to 3 ml must be drawn and discarded to free the specimen of heparin content. Heparin in the specimen could affect the results of the laboratory analysis.

Specimen Collection Procedure

A tourniquet is placed above the site and the plug tip wiped with alcohol. The 2-ml waste sample is withdrawn with a sterile needle and syringe through the butterfly plug. Another sterile needle and syringe are inserted and the blood specimen withdrawn before the tourniquet is released. Two milliliters of fresh sterile heparin solution are then injected into the line through the plug with a sterile needle and syringe, and the tubing is retaped securely to the patient's arm.

OBTAINING BLOOD SAMPLES FROM DIALYSIS AND KIDNEY TRANSPLANT PATIENTS

The venipuncturist drawing blood from patients in kidney dialysis and transplant units should be oriented before he attempts the procedure. The kidney patient can, and in some cases will, be on dialysis for a long time. The venipuncturist must be careful not to damage the veins. Often a cannula or fistula will be implanted in the patient's arm as an access for dialysis.

A red card on the door will indicate whether the dialysis or transplant patient has a cannula, fistula, or a gortex or bovine graft. The patient usually wears a yellow identa-band on the opposite arm from the cannula, fistula, or bovine or gortex graft to alert personnel of this condition. The patient's identification band is also put on the arm opposite the access.

OBTAINING BLOOD SPECIMENS FROM THE CANNULA

A cannula is used as an access for dialysis and for blood drawing. The cannula can be removed when no longer needed because of a successful transplant. The parts of a cannula system are

- the tips, which are inserted into the artery and vein;
- the cannula body tubes, which are made of smooth silicone rubber;
- the connectors, either a straight short Teflon connector or a "T" connector, which join the two body tubes and allow the blood to flow from artery to vein (*see* Fig. 1-36).

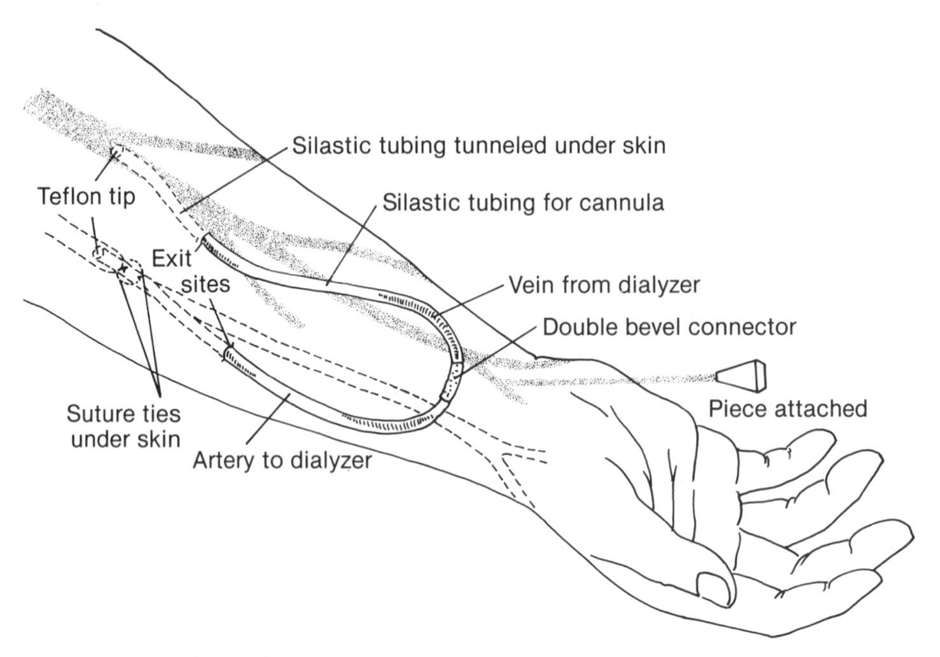

Fig. 1-36. *External cannula.*

For convenience in obtaining blood specimens from patients with cannulas, a "T" tube connector is used. On the external surface and in the middle of the "T" tube connector is a small tube about 15 cm (6 in) long. At the end of this small tubing is a rubber diaphragm cap, and in the middle of the cap is a small circle where the needle is inserted. The cannula with a "T" tube connector should not be used for drawing blood without the physician's permission. Drawing blood from the cannula with a "T" tube on kidney transplant or dialysis patients should be done only by specially trained personnel or by the transplant or dialysis teams. A cannula "T" tube is used for obtaining all blood specimens requested except those for blood cultures.

Objectives

- To maintain a sterile technique when drawing blood specimens from the "T" tube
- To maintain a safe technique for the cannula and "T" tube when drawing blood specimens
- To obtain blood specimens from a cannula "T" tube that will provide accurate blood studies

Procedure

A total of three syringes are used for this procedure.

1. Locate the rubber diaphragm cap under the Ace bandage.
2. Lift out the rubber diaphragm cap gently.
3. Wipe the rubber diaphragm cap with an alcohol sponge. Place the sponge under the cap.

4. Insert a 21-gauge or a 22-gauge needle with syringe in the small circle in the diaphragm. Draw out about 0.5 ml to 0.75 ml of blood and discard to clear the small tubing, since "T" tubes may be used for other purposes.
5. Wipe the rubber diaphragm cap with an alcohol sponge again. Using a new 21-gauge or 22-gauge needle and syringe, draw out the amount of blood needed for tests and disperse into proper tubes.
6. After the blood specimens are in the proper tubes, flush the "T" tube with sterile normal saline.
7. Wipe the normal saline vial with an alcohol sponge and draw out 0.75 ml of normal saline into a new needle and syringe.
8. Wipe the diaphragm cap with an alcohol sponge again. Insert the needle and syringe with normal saline and inject about 0.50 ml to 0.75 ml of solution.
9. Wipe the diaphragm cap with an alcohol sponge and tuck the cap under the Ace bandage.

Problems

1. If the rubber diaphragm cap or small tubing is clotted, no blood can be withdrawn. If the small tubing is free of clots, carefully guide a 21-gauge needle with syringe down the tube a short way and withdraw the needle to pull out the clot.

If this technique is used, be careful that the bevel of the needle does not pierce the small tube. If the "T" tube does get pierced and blood is oozing from the hole, clamp one of the cannula clamps on the small "T" tube. Call dialysis to alert them of the problem.

Another method is to flush the "T" tube with about 0.50 ml to 0.75 ml of sterile normal saline and, without taking the needle from the cap, withdraw about 1 ml of blood and discard. If this does not work, call dialysis for assistance.

2. Sometimes the rubber diaphragm cap will break off. Clamp a cannula clamp on the "T" tube. Alert dialysis to the problem.
3. Sometimes a break will occur where the small tube joins the connector and blood will ooze out. A cannula clamp on the small tube will not stop the oozing. Call dialysis immediately for assistance. Do not put cannula clamps on the cannula body tubes.

Do not insert a needle into the cannula body tubes for blood specimens because this could damage the cannula. When there are problems with cannulas, call dialysis for assistance.

FISTULA

The fistula, which in this case is a surgically fused vein and artery, is permanent and should never be used for drawing blood. Routinely, neither should the arm with the fistula be used for drawing blood. In patients with a fistula in one arm and an i.v. transfusion in the other, the arm with the i.v. line should not be used before checking with the patient's physician. If there are no alternative sites, the arm with the fistula can be used below the fistula. A hand vein is preferable; if that is not possible, the antecubital fossa area should be examined for a vein away from the fistula to the side or underside of the arm. A tourniquet may be applied loosely until a vein is located and then released as soon as possible. Pressure should be applied to the site after the needle has been withdrawn. Bleeding may occur because most of these patients are receiving heparin. A loose bandage may be applied.

The bovine graft, a piece of vein from an animal grafted to the patient's vein, and a Gore-Tex graft, a synthetic product, are sometimes used to fuse the vein and artery to develop a fistula. Usually the fistula must develop for 4 to 6 weeks before it can be used for dialysis.

THE ISOLATION ROOM

The Isolated Patient

Patients are isolated for the following reasons.

1. To contain the disease so that it is not spread to other patients, relatives, and employees.
2. To protect the patient from outside contamination. The patient's normal protective mechanism has been reduced to such a point that infection could be fatal.

The Isolation System

Isolated patients are assigned private rooms. A color-coded isolation sign is put on the door or window of the patient's room. The card describes the type of isolation and what precautions must be taken by those who enter the room.

There are eight types of isolation.

1. Strict (Yellow Card)

   Examples of diseases that require strict isolation are extensive burns, staphyloccocal pneumonia, chicken pox, disseminated herpes, rabies, and congenital rubella.

   The following precautions may be on the isolation door sign.

   - Private Room—necessary; door must be kept closed.
   - Gowns—must be worn by all persons who enter room.
   - Masks—must be worn by all persons who enter room.
   - Hands—must be washed on entering and leaving room.
   - Gloves—must be worn by all persons who have patient contact.
   - Articles—must be discarded or wrapped before being sent to Central Supply for disinfection or sterilization.

2. Respiratory (Red Card)

   Examples of diseases that require respiratory isolation are tuberculosis, either pulmonary suspected or sputum positive; meningococcal disease; Legionnaires' disease; influenza pneumonia; and measles, mumps, and rubella.

   The following precautions may be on the isolation door sign.

   - Private room—necessary; door must be kept closed.
   - Gowns—not necessary.
   - Masks—must be worn by all persons who enter room.
   - Hands—must be washed on entering and leaving room.

- Gloves—not necessary.
- Articles—those contaminated with secretions must be disinfected.
- Caution—all persons susceptible to the specific disease should be excluded from patient area; if contact is necessary, susceptible persons must wear masks.

3. Wound and Skin (Green Card)

Examples of diseases that require wound and skin isolation are moderate burns; purulent infections where drainage cannot be contained in simple dressing; puerperal sepsis; and eruptive herpes.

The following precautions may be on the isolation door sign.

- Private Room
- Gowns—must be worn by all persons who have direct contact with patient.
- Masks—not necessary, except during dressing changes and with burn patients.
- Hands—must be washed on entering and leaving room.
- Gloves—must be worn by all persons who have direct contact with infected area.
- Articles—special precautions necessary for instruments, dressings, and linen.

4. Protective (Blue Card)

Examples of conditions for protective isolation are agranulocytosis; immunosuppressive conditions; and in patients who receive immunosuppressive therapy.

The following precautions may be on the isolation door sign.

- Private Room—necessary; door must be kept closed.
- Gowns—must be worn by all persons who enter room.
- Masks—must be worn by all persons who enter room.
- Hands—must be washed on entering and leaving room.
- Gloves—may be worn by persons who have direct contact with patient.
- Articles—wash any dry tourniquet.

5. Enteric (Dark Brown Card)

Examples of diseases that require enteric precautions are acute diarrhea; shigellosis; salmonellosis; and those caused by campylobacter.

The following precautions may be on the isolation door sign.

- Private Room
- Gowns—must be worn by all persons who have direct contact with patients.
- Masks—optional.
- Hands—must be washed on entering and leaving room.
- Gloves—must be worn by all persons who have direct contact with patient or with articles contaminated with fecal material, urine, or blood.
- Articles—special precautions necessary for articles contaminated with urine, feces, or blood. Articles must be well cleansed and disinfected or incinerated.

6. Viral Hepatitis Precautions (Light Brown Card)

   Conditions requiring isolation for viral hepatitis precaution are acute viral hepatitis and hepatitis-B-positive findings.

   The following precautions may be on the isolation door sign.

   - Private Room
   - Needle and syringe disposal—special precautions.
   - Hands—must be washed on entering and leaving room.
   - Gloves and gowns—must be worn for direct contact with blood, urine, saliva, or feces.
   - Masks—not necessary.
   - Articles—special precautions necessary for articles contaminated with blood, urine, saliva, or feces. Articles must be disinfected or discarded.

7. Creutzfeld–Jacob Disease Precautions

   Condition requiring precaution for Creutzfeld–Jacob disease is suspected or diagnosed Creutzfeld–Jacob disease. Nurse should notify venipuncture and i.v. service.

   The following precautions may be on the isolation door.

   - Private Room—preferable.
   - Gowns—not necessary.
   - Masks—not necessary.
   - Hands—must be washed on entering and leaving room.
   - Gloves—must be worn by all persons who have direct contact with patient's CSF or blood. Gloves may be picked up at nurses' station.
   - Articles—special precautions necessary for articles contaminated with CSF and blood. Articles must be cleansed well and disinfected by sterilization.

8. Protective Cardiac ICU Isolation

   Some institutions have protective isolation for heart transplant patients; the sign on the door of the room reads "Protective Cardiac ICU Isolation." A special dress code is observed by all persons who enter the room.

General Information

Gowns, gloves, masks, and other supplies are always kept in the stand outside the patient's room. This "clean equipment area" must not become contaminated at any time. The new modern hospital often has an anteroom that serves as the "clean area;" in that setting the person entering the room can don a gown, mask, or gloves before entering the patient's room. Doctors' suitcoats and jackets are left here.

Isolation Room Procedure

1. Before entering the isolation room,

   - always read the isolation sign on the door, which will explain the type of isolation and what you must wear and do. WEAR WHAT YOU MUST—DO WHAT YOU MUST!
   - Check your orders and assemble the equipment needed for the patient.

Remember that anything taken into the room must be left there, discarded, or carefully cleansed if taken from the room.

- Take in the minimum equipment needed.
  Syringe and needle
  Tourniquet (there should be a tourniquet in each patient's bathroom, but it is wise to have one in your pocket)
  Proper tubes
  Four to five sponges (dry)
  Four to five sponges soaked with alcohol
  Glass slides (if necessary)
  Small roll bandages (when finished, leave leftover roll, if any, in patient's bathroom)

2. On entering the room,
   - place four or five paper towels from bathroom on table and place equipment on one or two towels spread open.
   - wash hands.
   - in protective isolation (blue card), wash and dry tourniquet. Specific specimen handling is not necessary, such as the use of paper towels and holding tubes with dry sponges. In all other types of isolation, however, the procedural steps below are recommended.
   - obtain blood samples in the usual manner using a syringe, avoiding any unnecessary contact with the patient and bed.
   - remove a paper towel from top of supply on table, lay syringe and needle on it while you bandage the arm.
   - pick up the tubes to be filled with dry sponges in left hand (right hand, if left handed).
   - fill tubes, holding syringe and needle in right or left hand.
   - lay filled tubes on clean paper towel.
   - discard syringe and needle in receptacle in bathroom.
   - remove gown and dispose of in proper container.
   - wash hands and tourniquet with soap and water and dry.
   - turn faucet off with clean paper towel so as not to contaminate hands.
   - pick up tubes from paper towel, clean with alcohol pad, and leave patient's room.
   - if patient has hepatitis, label tubes with "H" label. (These tubes should never be sent through pneumatic tube unless in a plastic bag.)
   - if blood smears were made, place smears on two clean paper towels. When ready to leave, wrap smears and tubes in top paper towel and discard bottom paper towel.
   - never take trays and carts in isolation rooms.

3. The venipuncture technician should be aware of the following situations.
   - Pregnant venipuncturists should not enter an isolation room when the patient has rubella or cytomegalovirus infection. The hospital nurse–epidemiologist should inform the clinical laboratory when rubella or cytomegalovirus is suspected. A patient suspected of having cytomegalovirus but not in isolation should have a notice posted on the door card that will alert the venipuncturist.

- Gloves should be worn when dealing with patients isolated for burns or wounds.
- Accidental needle prick or contamination of a break in the skin by blood or excreta from an infected patient should be reported immediately to the employee health service.

COLLECTION TECHNIQUES FOR SPECIAL TESTS

BLOOD ALCOHOL TEST

When a specimen has been ordered for a blood alcohol test, the venipuncture site should not be cleaned with alcohol because this may contaminate the specimen, resulting in falsely elevated results. The site may be cleaned with regular soap and water but must be completely dry before puncture is attempted.

Usually, a venipuncturist draws a specimen for blood alcohol only when the patient's physician has ordered the test for medical reasons. At most institutions a different procedure is followed if a blood alcohol test is ordered for legal reasons from the Police Department, the Highway Patrol, or the Sheriff's Office. In this situation, a physician draws the blood, which is then put in a special container and given to the law-enforcement officer.

COAGULATION TESTS

Because diagnostic and therapeutic decisions are based on coagulation test findings, it is important to have documented procedures for collecting specimens for special coagulation studies. Many variables such as the type of anticoagulant, drawing procedure, and the storage of the sample may affect the final test result.

Blood specimens for coagulation studies should not be collected in containers made from soda lime or soft glass, but rather in tubes made from nonreactive materials that will not interact with the coagulation system, such as plastic, borosilicate glass, or siliconized glass.

The anticoagulant of choice is usually 109 or 129 mM/liter (3.2% or 3.8%) of sodium citrate solution (the dehydrate form): $Na_3C_6H_5O_7\text{-}2H_2O$. Buffered citrate solutions are also acceptable anticoagulants for obtaining blood for coagulation tests. The final concentration with the dehydrate sodium citrate should be 10.9 to 12.9 mM/liter in the blood sample. Some special studies require oxalates as the anticoagulant.[8]

Poor venipuncture techniques may contribute to false results in coagulation studies. The activated partial thromboplastin time (APTT) is particularly prone to variations caused by poor venipuncture technique. An "atraumatic" venipuncture is imperative because introducing tissue and tissue juices into the specimen may result in falsely abnormal coagulation results.

Ideally, blood specimens for coagulation testing should be drawn individually rather than as part of a large amount of blood for other tests. Because this is often impractical, however, it is acceptable for the sample to be part of a blood specimen of no more than 25 ml.

The two-syringe technique using the blood from the second syringe is preferable for most coagulation procedures. Blood should be obtained with a plastic syringe (no larger than 35 ml) and a 19-gauge needle. Undue suction in filling the syringe may result in hemolysis. Probing, air bubble aspiration, hematomas, contamination with tissue fluids, and prolonged venous stasis are complications to avoid.

Immediately after the venipuncture has been completed, the needle should be removed from the syringe and the blood allowed to run down the side of the tube with the anticoagulant, never squirted. The ratio of anticoagulant to blood is critical; the venipuncturist must be sure that the volume ratios are correct. All covers and caps should be removed from the tubes before drawing blood so that the blood can be placed immediately into the anticoagulant. The tube should then be covered and mixed immediately by gentle inversion of the tubes three or four times. Shaking the tube or producing bubbles must be avoided.

In an evacuated tube system, the coagulation specimen should be the second or third tube. If the blood specimen is drawn from an indwelling catheter, the first 20 ml of blood must be discarded or used for other tests before the specimen for coagulation testing is obtained. If the line contains heparin, at least 30 ml of blood must be discarded or used for other tests before the coagulation specimen is drawn.

TIMED INTERVALS

Some specimens must be drawn at timed intervals because of medications or biologic variations (circadian rhythms). Specimens must be obtained at the precisely specified interval, and specific directions to that effect must be given to the venipuncture team. The following tests require timed specimens.

- Tests where diurnal or other time effects may be anticipated, such as corticosteroid, serum iron, and glucose tolerance.
- Therapeutic-drug-monitoring tests, such as prothrombin time, salicylic acid, digoxin, and tricyclic antidepressants.

In these cases, the time of the specimen collection should be recorded accurately on the request slip or in the laboratory computer system.

TRACE METAL ANALYSES

Two syringes are used to obtain specimens for trace metal analysis. One, the Sarstedt, must be thoroughly cleaned by leaching in 15% nitric acid for a week and dried in a class 100 atmosphere. This syringe is used to obtain the specimen sent to the laboratory.[9] The second syringe, which has not been specially prepared for obtaining samples for trace element analysis, can be a disposable syringe normally used to obtain specimens for other analyses. The procedures for using the syringes for obtaining blood specimens for trace metal analyses are as follows.

- Place a stainless needle with a plastic hub on a disposable syringe that has not been subjected to the elaborate cleaning process, and position the needle in the vein.
- Draw 2 ml to 4 ml of blood into this syringe at a relatively rapid rate, remove the syringe from the needle, and discard the syringe and contents. (If blood specimens are to be obtained at the same time for other tests, the first syringe can be used for that purpose, and further rinsing of the needle is not necessary.)
- Place a cleaned Sarstedt syringe on the needle and slowly draw 10 ml of blood.
- Replace the needle on the syringe with the plastic cap, unscrew the plunger rod from the syringe, and send the Sarstedt syringe that contains the specimen to the laboratory.

Trace metals include the following.

Lead (Pb) Mercury (Hg)	Draw in a Sarstedt syringe with EDTA.
Arsenic (As) Aluminum (Al) Chromium (Cr) Cadmium (Cd) Manganese (Mn) Zinc (Zn)	Draw in a plain Sarstedt syringe.

REFERENCES

1. Risse JB: The renaissance of bloodletting: A chapter in modern therapeutics. J Hist Med Allied Sci 34:3, 1979
2. Sergworth Jr: Bloodletting Over the Centuries, NY State J Med 13:2022, 1980
3. National Committee for Clinical Laboratory Standards. Standard Procedures for the Collection of Diagnostic Blood Specimens by Venipuncture. Adopted Standard: H3-A. NCCLS, 771 East Lancaster Avenue, Villanova, PA 1980
4. National Committee for Clinical Laboratory Standards. Standard for Evacuated Tubes for Blood Specimen Collection. Adopted Standard H1-A. NCCLS, 771 East Lancaster Avenue, Villanova, PA 1980
5. Jones JD: Factors that Affect Clinical Laboratory Values. J Occup Med 22:316, 1980
6. Calum R, Cooper MH: Letter to Editor, Recommended "Order of Draw" for Collecting Blood Specimens into Additive Containing Tubes. Clinical Chemistry 28: 1399, 1982
7. Ladenson JH: Gradwhols' Clinical Laboratory Methods, 8th ed, CV Mosby Co., St. Louis Missouri, 1980
8. National Committee for Clinical Laboratory Standards. Guidelines for the Collection, Transport and Preparation of Blood Specimens for Coagulation Testing. Proposed Standard H21-P, p 9, NCCLS, 771 East Lancaster Avenue, Villanova, PA, 1981
9. Moody JR, Lindstrom RN: Selection and cleaning of plastic containers for storage of trace element samples. Anal Chem 49:2264-2267, 1977

SUGGESTED READING

Avery GB: What pediatric patients need from laboratories. Clin Chem News 6(11):7S, 1980

Becton–Dickinson and Company: Seminar Stresses Importance of Specimen Collection. Rutherford, New Jersey, Becton–Dickinson Vacutainer Systems, 1981

Bedford RF: Long-term radial artery cannulation: effects on subsequent vessel function. Crit Care Med 6:64–67, 1978

Bjueletich J, Hickman RO: The Hickman Indwelling Catheter. J Nurs 80:62–65, 1980

Blackburn EK et al: Blood specimen collection tubes for coagulation tests. J Clin Pathol 32:741, 1979

Bremer R: We draw multiple specimens without multiple sticks. MLO, p 103, November 1976

Brown BA: Hematology: Principles and Procedures. Philadelphia, Lea and Febiger, 1973

Brown HI: Lectures for Medical Technologists. Springfield, Illinois, Charles C Thomas, 1964

Calam RR: Reviewing the importance of specimen collection. J Am Med Tech 39:297, 1977

College of American Pathologists: *So You're Going to Collect a Blood Specimen.* Danville, Illinois, Interstate Printers and Publishing Co, 1974

Committee on Standards, American Association of Blood Banks: Standards for Blood Banks and Transfusion Services, 10th ed. Washington DC, American Association of Blood Banks, 1981

Desquitado M, Huestis B: How we reduce our bad specimen problems. Lab Management, p 24, February 1977

Dito WR: Theraputic drug monitoring in your laboratory. Diag Med 3:21–25, 1980

Drosdowsky M et al: Factors to be taken into consideration for blood sampling in view of establishing reference values (document E, stage 3). Ann Biol Clin (Paris) 38(4)251–265, 1980

DuBois JA: *Course in print 6-Blood circulation and venipuncture.* Lab World 31:22–28, July 1980

Franz T: For phlebotomists: ending the collection tube confusion. MLO 10:157, 1978

McNair P et al: Gross errors made by routine blood sampling from two sites using a tourniquet applied at different positions, Clin Chim Acta 98:113–118, 1979

McPhedran P et al: Prolongation of the activated partial thromboplastin time associated with poor venipuncture technique. Am J Clin Pathol 62:16–20, 1974

Palermo LM et al: Avoidance of heparin contamination in coagulation studies drawn from indwelling lines. Anesth Analg (Cleve) 59:222–224, 1980

Slockbower JM: Venipuncture procedures. Lab Med 10:747–752, 1979

Skin Puncture Blood-Specimen-Collection

Thomas A. Blumenfeld

Recently the number of blood samples collected from ill and well children for diagnostic laboratory tests has increased greatly, primarily owing to more ill newborns being cared for in intensive-care units and new legally mandated screening programs to detect metabolic defects in newborns. When blood specimens are obtained from children for diagnostic laboratory tests, avoidance of injury from collecting too large a specimen or from the method of collection is paramount.

The blood volume of young children, especially premature newborns, may be very small (Fig. 2-1),[1] and if blood samples for diagnostic tests are taken without considering a patient's size or the frequency of blood removal, a hospital-induced anemia may occur in addition to the child's other medical problems. For this reason, a daily log should be kept for each hospitalized child to record the amount of blood and the time of day that each blood specimen is obtained.

Because the quantity of blood removed and the avoidance of injury during specimen collection are important considerations, blood collection by skin puncture is the technique of choice for collecting small amounts of blood from children, especially newborns. In children, depending primarily on age, skin-puncture blood may be obtained from the heel or the distal phalanx of a finger.

HEEL PUNCTURE

Heel puncture is generally performed in children younger than 1 year of age; after this age they begin to stand and walk. The skin puncture site must not be swollen; a swollen site indicates that tissue fluid or blood has accumulated within the skin, and the addition of these to the blood specimen may cause erroneous results in diagnostic tests. With an infant's heel, to avoid puncturing the underlying bone (calcaneous) and the risk of osteochondritis, the veni-

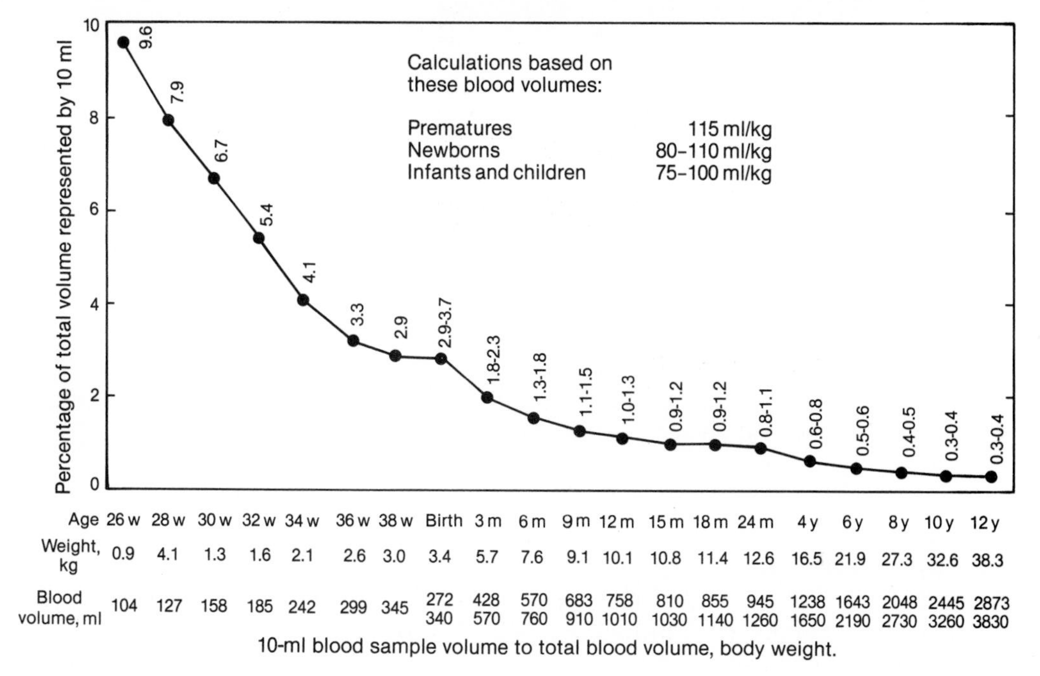

Fig. 2-1. *Relationship of 10-ml of blood sample volume to total blood volume, body weight, and age of infant or child.*

puncturist should follow these general guidelines based on anatomical relationships.[2]

1. Puncture on the most medial or lateral portions of the plantar surface of the heel, medial to a line drawn posteriorly from the mid-great toe to the heel or lateral to a line drawn posteriorly from between the fourth and fifth toes to the heel (Fig. 2-2).
2. Puncture no deeper than 2.4 mm.
3. Do not puncture on the posterior curvature of the heel because the distance from the skin to the bone is one half that from the plantar surface to the bone.
4. Do not puncture through previous puncture sites that may be infected.

Infant heel puncture should be no deeper than 2.4 mm because, in small premature infants, the heel bone may be as close as 2.4 mm beneath the plantar skin surface, and in all newborns the skin's major blood vessels are located at the dermal subcutaneous junction 0.35 mm to 1.6 mm beneath the skin surface. Thus a skin puncture anywhere on the heel of an infant, regardless of age, need not be deeper than 1.6 mm to reach the appropriate blood vessels but could go as deep as 2.4 mm and not strike the bone. Commercially available skin puncture lancets have blades of various designs, length, and width,[3] some of these are longer than 2.4 mm and may be hazardous if used to puncture heels of premature newborns. For full-term newborns and older children, however, if the venipuncturist follows all guidelines except for the 2.4 mm depth,

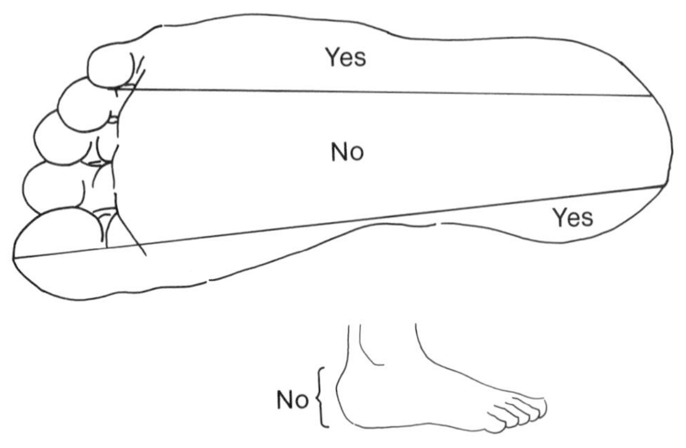

Fig. 2-2. *Recommendations for heel skin punctures in newborns:* [1] *Perform punctures on the most medial or most lateral portion of the plantar surface (outside the lines on the diagram);* [2] *puncture no deeper than 2.4 mm;* [3] *do not perform punctures on the posterior curvature of the heel; and* [4] *do not puncture through previous sites that may be infected. (Blumenfeld TA, Turi GK, Blanc WA: Recommended sites and depth of newborn heel skin punctures based on anatomic measurements and histopathology. Lancet 1:213, 1979)*

the chance of bone puncture will be remote, and lancet blades equal to, or less than, 5 mm can be used for two reasons: [1] In 95% of newborns, the heel bone does not extend medial to a line drawn posteriorly from the mid-great toe to the heel or lateral to a line drawn posteriorly from between the fourth and fifth toes to the heel; if the punctures are performed outside these lines, they can be deeper than 2.4 mm, and the risk of striking the bone is negligible; and [2] in term infants, the heel bone is 5 mm, or more, beneath the skin. In term newborns and older infants, the lancets currently available that have blades equal to, or less than, 2.4 mm in length frequently produce punctures that yield very small specimens and require the phlebotomist to make more than one puncture to obtain the needed blood volume for testing. Blades equal to, or less than, 2.4 mm in length may produce punctures that yield insufficient blood primarily because they are quite narrow. Future lancets should be designed with a blade wide enough to cause an adequate blood flow but short enough to prevent the risk of bone penetration.

In the past, Bard–Parker #11 surgical blades have been used to puncture the skin for blood samples. This blade was originally designed to lance abscesses and to create a surgical incision. It can also create wounds that are much deeper and larger than needed for blood collection and should *not* be used for skin punctures because it is too hazardous.

Sometimes skin punctures in newborn feet are performed in the arch of the foot: The rationale is to puncture the lateral or the medial plantar artery and vein that traverse this area. The results of an anatomical study of the arch in newborn feet indicate that skin punctures for obtaining blood specimens should not be performed there because they may injure nerves, tendons, and cartilage and offer no advantage over punctures in the skin of the heel.[4]

FINGER PUNCTURE

Skin punctures on the palmar surface of the distal phalanx of the finger should not be performed in infants, especially premature ones. In newborns, the distance from the skin surface to the underlying bone in all fingers varies from 1.2 mm to 2.2 mm, and with the available lancets the bone in newborn fingers could be punctured, possibly resulting in osteochondritis.[5] Local infection and gangrene may also be complications of finger punctures in this group.[6] Finger punctures may be performed on children older than 18 months or on adults, and these guidelines should be observed.[5]

1. Perform finger punctures on the center of the distal phalanx on the palmar surface; do not perform them on the side or tip of the finger because the tissue thickness in these areas is about one half of that in the center of the finger.
2. Do not puncture deeper than 3.1 mm because the distance from the skin surface to the bone may vary from 3.1 mm to 10.9 mm.

PREPARATION OF PUNCTURE SITE

Before puncturing the skin, the venipuncturist should warm the site to increase the blood flow to the area as much as sevenfold.[7] Because the increase is primarily in arterial blood flow, the blood specimen obtained after warming is called *arterialized skin-puncture blood.* This warming step is essential for accurate results when specimens are collected for *p*H and blood gas determinations.[8] The simplest and least expensive method of warming is to cover the site for 3 minutes with a hot, moist towel at a temperature no higher than 42°C, which can be achieved by placing a cloth towel under hot tap water until it is hot but does not cause great discomfort to the person holding it. This warming technique adequately increases blood flow but does not burn the skin or result in significant changes in the chemical values routinely measured in hosital chemistry laboratories.[9]

The chosen skin puncture site should be cleaned with 70% aqueous solution of isopropanol (70% V/V). After cleaning, the skin must be completely dried with a sterile gauze pad before puncture because any remaining alcohol will cause rapid hemolysis of blood that touches it. Povidone iodine (Betadine) should not be used to clean skin puncture sites because blood contaminated with it may have falsely elevated levels of potassium, phosphorus, uric acid, and bilirubin.[10]

COLLECTION TUBES

Skin-puncture blood may be collected by capillary action into a capillary tube—the preferred way—or drop by drop into a small test tube, which increases hemolysis.[11] Capillary tubes with different bore sizes and capacities are available,[1] and the use of heparinized or nonheparinized tubes depends on whether plasma or serum is desired.

A relatively new product for skin puncture specimen collection consists of a small plastic tube with a capillary tube as part of the lid,[12] which allows the specimen to be collected by capillary action and the user to have a large specimen in a single container. After collection, the lid that contains the capillary tube is removed and the container capped and centrifuged. Some tubes contain a serum separater, and after centrifugation the serum can be poured directly from the tube.

The advantages of collecting skin-puncture blood in small-bore capillary tubes are that the specimen can be collected in several tubes and distributed to different laboratory work areas for different tests; and, if one tube breaks or contains a hemolyzed specimen, other tubes are available. The advantages of using large-bore capillary tubes is that after the specimen is collected and centrifuged, the tube need not be broken to obtain the serum or plasma, and the specimen may then be pipetted from the tube's open end into an analysis cup. The disadvantage of the larger capillary tube is that if one breaks or contains hemolyzed blood, a large portion of the specimen may be lost or unusable.

BLOOD COLLECTION TECHNIQUE

After the chosen site has been prepared, the venipuncturist should grasp the infant's heel with a moderately firm grip, with the index finger at the arch of the foot and the thumb placed proximal to the puncture site at the ankle. The skin should be punctured at a slight angle to the skin surface. After the puncture, the first drop of blood should be wiped away with sterile gauze or cotton because it is most likely to contain tissue fluids that may contaminate the specimens. The blood flowing from the puncture site will form a drop over the site, and, when a capillary tube tip is placed against the drop, blood will flow into the tube by capillary action. Blood flow from the site will be increased if the puncture site is held downward and a gentle continuous pressure applied to the surrounding tissue. Strong repetitive pressure (milking) should not be used because it may cause hemolysis and increase the amount of tissue fluid in the specimen. If an adequate puncture has been done, 0.5 ml to 1 ml of blood can be collected from a single puncture site.

After blood has been collected, the infant's foot should be elevated above the rest of his body and a sterile gauze pad or cotton swab pressed against the puncture site until the bleeding stops. A bandage over the puncture site may be a hazard: An adhesive bandage may irritate the infant's skin, and infants and young children may remove the bandage, put it in their mouths, and possibly aspirate it. Thus it is best not to apply adhesive bandages over skin puncture sites in infants younger than 2 years of age.[3]

When skin-puncture blood is collected for *p*H and blood gas determinations, the site must be warmed adequately before puncture to obtain "arterialized blood."[8] The specimen should be collected in heparinized glass capillary tubes and not contain air bubbles. Blood specimens exposed to room air for as short as 10 seconds to 30 seconds can have significant errors in Po_2 determinations. Bubbles of room air in the specimen cause the original PO_2 in the specimen to move toward that of the room air PO_2. The air bubbles in the specimen have little effect on the specimen's PO_2 until the surface area of the blood exposed to the air bubble is increased by mixing.[13] Because whole blood specimens in capillary tubes must be mixed before *p*H and blood gas analyses, careful technique during specimen collection must be used to avoid air bubbles in the capillary tubes.

Capillary tubes can be sealed with commercially available sealant or caps made for this purpose. If a specimen is collected for routine clinical chemistry tests, only one end of the tube need be sealed. When capillary tubes are sealed with sealant, they should be held at about 45° and inserted into the material with a scooping motion, which will force it into the tube. The tube should be inserted 3 mm to 4 mm deep into the sealant and rotated between the thumb and forefinger before being removed to assure that the sealing material will remain in the bore of the tube after the tube is withdrawn.

Specimens collected for *p*H and blood gas analyses should have one end of the tube sealed immediately with sealing material and a small magnetic mixing bar placed in the bore of the tube, and then the opposite end sealed. With the mixing bar in the tube, the blood can be mixed by moving a magnet back and forth along the outside of the tube before analysis. During transportation to the laboratory, specimens for *p*H and blood gas analyses should be placed in water that contains ice chips to prevent a significant change in *p*H. The *p*H of whole blood will not change significantly for 4 hours if the specimen is immersed in water that contains ice. At 27°C, however, the *p*H will change about 0.005 units every 10 minutes, and at 37°C the change will be twice this amount.[14] The *p*H decay rate is independent of the hemoglobin concentration but very dependent on the leukocyte count.[15] Even in specimens from leukemic patients with elevated leukocyte counts, if specimen containers are placed in ice water the *p*H decay rate is no greater than 0.0008 *p*H per hour.[16] The stability of PO_2 is also very temperature dependent, and blood samples for this measurement should be chilled during transport.[13]

Skin puncture blood-specimen containers such as capillary tubes and small test tubes are usually too small for each container to be labeled separately. To prevent identification errors, several small sealed tubes may be wrapped with one adhesive label (like a flag) or several from the same patient may be placed in one labeled large tube. The labels should contain the patient's name, hospital number, hospital location, and date and time of collection.

SKIN PUNCTURE COMPLICATIONS

Complications of skin-puncture blood collected by heel punctures from young children are osteochondritis of the heel bone (calcaneous) and microabscesses of the overlying skin; both can be prevented by proper blood collection techniques.[2] Another complication is foci of calcification of the heel skin of newborns who have numerous heel punctures,[17] a condition thought to result from foci of epidermis being displaced into the underlying tissue. It appears at 4 to 12 months, disappears at 18 to 30 months, and is almost always asymptomatic. In only one instance did a child have signs of discomfort while wearing shoes.[17]

Reported complications of skin punctures of the distal phalanx of the fingers of newborns are puncture-site infection and gangrene.[6]

COMPARISON OF CHEMICAL CONSTITUENTS USING DIFFERENT TECHNIQUES

Blood obtained by skin puncture is a mixture of blood from arterioles, venules, and capillaries and contains interstitial and intracellular fluids. Because of its unique composition, specimens obtained by skin puncture should specifically be called skin-puncture blood, not capillary blood.

Blumenfeld and colleagues compared 12 chemical values in simultaneously obtained skin-puncture serum, skin-puncture plasma, and venous serum and studied the effect of warming the skin-puncture site before the puncture.[9] They found no clinically important differences in the concentration of the chemicals measured in skin-puncture serum and plasma with or without warming the skin before puncture. When each measured concentration was compared in skin-puncture blood and venous serum, however, the concentrations of glucose, potassium, total protein, and calcium had differences that could be clinically significant. Except for glucose, the concentration in venous serum of each of these was higher. The degree of hemolysis reflected by the measurement of free hemoglobin was the same in skin-puncture serum and skin-puncture plasma but much greater (50–60%) than in venous serum. This greater free hemoglobin, indicating hemolysis in skin-puncture specimens, was not reflected, however, in clinically significant increases in lactic dehydrogenase and potassium values. Blumenfeld and coworkers concluded that there are differences in some chemical concentrations in skin-puncture specimens and venous serum and that these do not limit the usefulness of skin-puncture specimens but should be considered when results from these specimens are compared.[9] The study was performed on healthy adult volunteers; the results probably would be no different in children except for the concentrations of free hemoglobin in newborn skin-puncture blood, which is higher than in adults.[11]

When *p*H and P_{CO_2} in arterialized skin-puncture blood and arterial blood from newborns are compared during the first hour of life, the skin-puncture samples reliably reflect arterial values in 40% of healthy newborns. After 3 hours of life there is agreement in 90% of healthy newborns. In newborns with abnormal heart and lung function, however, the agreement is present in only 70% of cases.[8] Therefore, arterialized skin-puncture blood specimens may not be reliable for measuring *p*H or P_{CO_2} in the first hour of life or in newborns with heart and lung disease.

The P_{O_2} in arterialized skin-puncture blood and umbilical artery blood obtained during the first few days of life, correlates poorly in newborns with various illnesses or those who are premature, and almost uniformly P_{O_2} is lower in skin-puncture blood than in arterial blood.[18] Thus the arterial P_{O_2} tension may be at toxic levels but remain undetected in arterialized skin-puncture samples; results from an umbilical arterial catheter sample or a transcutaneous oxygen electrode should be used to monitor arterial P_{O_2}. The *p*H and P_{CO_2} in newborns with abnormal heart and lung function should be measured using arterial blood.

Differences in blood cell counts and coagulation factors in arterial blood, venous blood, and skin-puncture blood of newborns have been studied. In healthy term newborns, leukocyte counts were made in arterial blood, venous blood, and skin-puncture blood obtained simultaneously during rest, mild exercise, and violent exercise.[19] At rest, the umbilical vein total leukocyte counts were 82% below skin-puncture values, and the umbilical arterial leukocyte counts were 77% below skin-puncture value. When the infants were crying violently, the skin-puncture leukocyte counts increased 146% above base line and immature neutrophils increased. During mild exercise, skin-puncture leukocyte counts increased 113% above base line but the neutrophils did not change. These elevated counts returned to base line in 1 hour. We therefore recommend that serial leukocyte counts be obtained from a consistent vascular source and from resting infants. If the infant has been crying violently, 1 hour should elapse before a blood sample is taken for a leukocyte count.

Platelet counts in venous blood and skin puncture blood in thrombocytopenic children show that the platelet counts in skin-puncture specimens are significantly lower than in venous specimens. The difference is most marked in newborns.[20]

Venous and skin-puncture blood from healthy and ill newborn infants showed no significant differences when tested for platelet counts, prothrombin–proconvertin, serum–fibrin degradation products, fibrinogen, and factor 5.[21] When blood is collected for coagulation studies in newborns, the technique of collection is probably more important than the source of blood.

REFERENCES

1. Blumenfeld TA: Clinical application of microchemistry. In Werner M (ed): Micro-Techniques for the Clinical Laboratory: Concept and Application, pp 1–15. New York, John Wiley & Sons, 1976
2. Blumenfeld TA, Turi GK, Blanc WA: Recommended sites and depth of newborn heel skin punctures based on anatomic measurements and histopathology. Lancet 1:213, 1979
3. Meites S, Levitt MS, Blumenfeld TA, Hammond KB, Hicks JM, Hill GJ, Sherwin JE, Smith EK: Skin puncture and blood collecting technique for infants. Clin Chem 25:183–189, 1979
4. Blumenfeld TA: Skin punctures in the arch of the feet of newborns. Unpublished data, 1981
5. Blumenfeld TA: Skin punctures for blood specimens from adults and infants fingers. Unpublished data, 1981
6. Karna P, Poland RL: Monitoring critically ill newborn infants with digital capillary blood samples: an alternative. J Pediatr 92:270–273, 1978
7. Guyton AC: Textbook of Medical Physiology, 4th ed, pp 206, 375. Philadelphia, WB Saunders, 1971
8. Gandy G, Grann L, Cunningham N, Adamson K Jr, James LS: The validity of pH and pCO_2 measurements in capillary samples in sick and healthy newborn infants. Pediatrics 34:192–197, 1964
9. Blumenfeld TA, Hertelendy WG, Ford SH: Simultaneously obtained skin puncture serum, skin puncture plasma and venous serum compared and effects of warming the skin before puncture. Clin Chem 23:1705, 1977
10. Van Steirtegham AC, Young DS: Povidone-iodine (Betadine) disinfectant as a source of error, Clin Chem 23:1512, 1977
11. Michaelsson M, Sjolin S: Hemolysis in blood samples from newborn infants. Acta Pediatr Scand 54:325–330, 1965

12. Hicks JR, Rowland GL, Buffone GJ: Evaluation of a new blood collection device (microtainer) that is suited for pediatric use. Clin Chem 22:2034–2036, 1976
13. Ladenson JH: Non-analytical sources of laboratory error in pH and blood gas analyses. In Durst RA (ed): Blood pH, gasses, and electrolytes, pp 175–190. Washington, DC, US Government Printing Office, 1976. National Bureau of Standards Special Publication 450
14. Gambino SR: Blood pH, pCO_2, oxygen saturation and pO_2. Chicago, American Society of Clinical Pathologists Commission on Continuing Education, 1967
15. Kelman GR, Nunn JF: Nomograms for correction of blood PO_2, PCO_2, pH and base excess for time and temperature. J Appl Physiol 21:1484, 1966
16. Gambino SR, Astrup P, Bates RG, Campbell EJM, Chinard FP, Nahos GG, Siggard–Anderson O, Winters R: Report of the Ad hoc Committee on Acid–Base Methodology. Am J Clin Pathol 46:376, 1966
17. Sell EJ, Hansen RC, Struck–Pierce S: Calcified nodules on the heel: A complication of neonatal intensive care. J Pediatr 96:473–475, 1980
18. Mountain KR, Campbell DG: Reliability of oxygen tension measurements on arterialized capillary blood in the newborn. Arch Dis Child 45:134–138, 1970
19. Christensen RD, Rothstein G: Pitfalls in the interpretation of leukocyte counts of newborn infants. Am J Clin Pathol 72:608–611, 1979
20. Feusner JH, Behrens JH, Detter JC, Cullen TC: Platelet counts in capillary blood. Am J Clin Pathol 72:410–414, 1979
21. Stuart J, Picken AM, Breeze GR, Wood BSB: Capillary blood coagulation profile in the newborn. Lancet 2:1467–1471, 1979

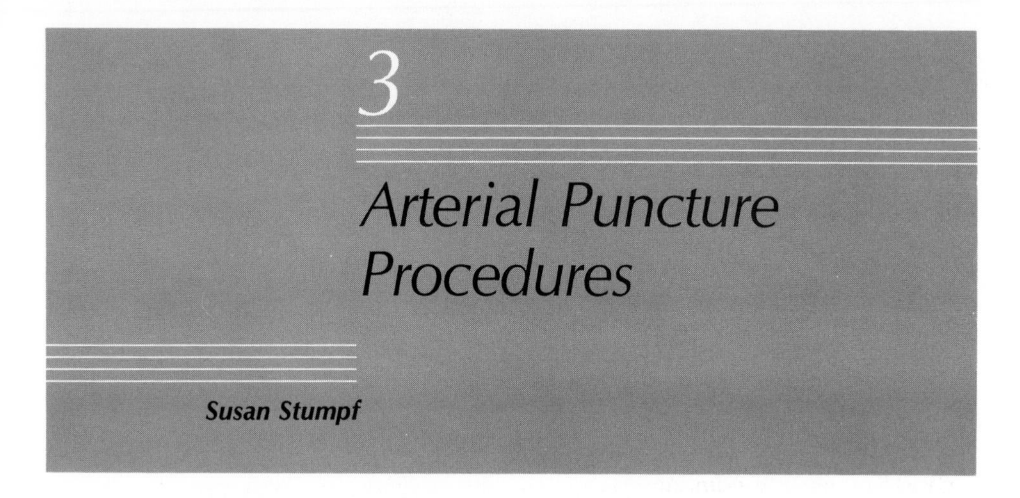

Arterial puncture was once considered dangerous and was reserved for only the most extreme circumstances. The current attitude toward this medical practice is very different. Recent automation of blood-gas analysis instruments has allowed faster analysis, more accuracy, and smaller sample size. Blood gas analysis has become more valuable in evaluating respiratory-related medical problems.

Because of increased demands on the physician's time and the increased use of blood gas analysis, paramedical personnel are now performing this service when needed. Paramedical personnel must go through a strict disciplined training program (*see* Chap. 12).

ESSENTIALS OF BLOOD GAS SAMPLES

An arterial blood gas specimen for analysis is of value to the physician only if

- the patient is prepared properly;
- the sample is collected properly;
- the correct amount of sample is actually obtained in relation to the size of the syringe; and
- the specimen is handled properly.

Persons performing arterial punctures should be familiar with the dangers of the procedure and with the precautions designed to prevent injury to the patient or alteration of the results.

Blood gas analysis measures the pressure exerted by the gases that we inhale and exhale dissolved in blood, including Po_2—the pressure of dissolved oxygen exerted in the blood—and Pco_2—the pressure of dissolved carbon dioxide exerted in the blood. The *p*H, a measure of the acid–base balance in the blood, is also measured in blood gas analysis. In the human body a *p*H of 7.40 is a perfect balance of acids and bases.

Arterial blood, which supplies each body organ's metabolic needs, is usually uniform throughout the body, whereas venous blood varies in composition depending on the size and activity of the tissue it has bathed. The greatest difference between arterial and venous blood is its oxygen content, but *p*H and carbon dioxide content also vary.

Because Po_2, Pco_2, and *p*H are so transient *in vivo,* the patient *must* be prepared properly. For accuracy, the patient must be in a steady-state condition, attained by having the patient rest for 20 to 30 minutes. Although the Po_2 and Pco_2 achieve their levels in the blood in only about 10 minutes, hemoglobin takes 20 to 30 minutes to saturate to its fullest with oxygen. Exertion can cause the hemoglobin to desaturate, thus altering the results. Changes in breathing patterns cause alterations in results in less than 1 minute. Examples of such situations include eating, experiencing pain, apprehension, breath holding, coughing, and exertion.

Once the patient is readied, the appropriate personnel should be notified and the sampling time scheduled. An arterial blood gas sampling should be scheduled no more than 1 hour in advance.

Some situations, including cardiac arrest, cardiac arrhythmias, respiratory arrest, atelecteosis (collapsed lung), and pulmonary emboli, effect a drastic change in a patient's condition. The sudden onset of these will require an arterial blood gas analysis immediately for the physician to evaluate accurately the patient's respiratory and metabolic status.

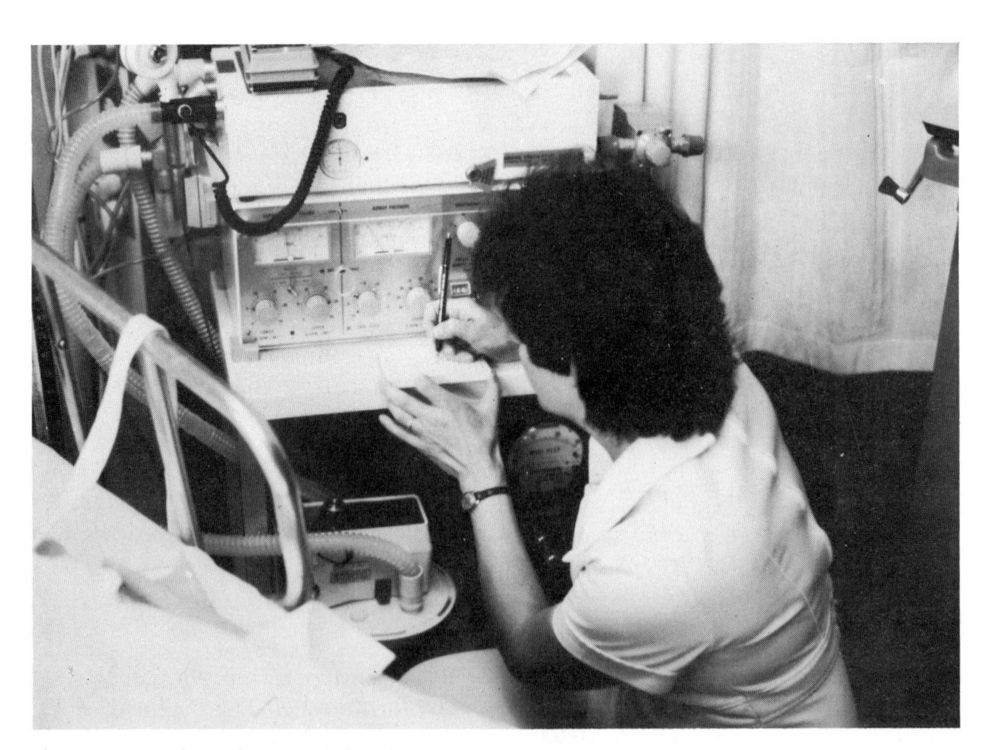

Fig. 3-1. *Recording of essential data from patient's ventilator.*

PATIENT DATA

Accurate arterial blood gas analysis involves recording the patient's condition at the time the sample is obtained. Included should be

- sampling time;
- FIO_2 (fraction of inspired oxygen;
- mode of delivery for oxygen therapy;
- patient's respiratory rate, including any significant pattern of breathing (*i.e.*, Cheyne-Stokes);
- site where the blood was drawn;
- the patient's body temperature;
- ventilator setting if the patient is on a ventilator (Fig. 3-1), including
    - tidal volume,
    - peak water pressure,
    - end expiratory pressure, and
    - mechanical versus spontaneous respirations.

The established procedure for arterial puncture must be followed strictly to maintain the quality of arterial samples. Discipline is critical. The quality of specimens should be monitored closely. Poor technique and sampling must be corrected immediately.

SAMPLING FACTORS THAT ALTER RESULTS

HEPARIN DILUTION

Sodium heparin is the most widely used anticoagulant for blood gas sampling because a minimal amount will effect anticoagulation of the greatest volume of blood with the least effect on acid–base values. When a liquid form is used for blood gas analysis, heparin should be found only in the dead space and the needle of the syringe for accurate results. Excess heparin can greatly acidify a blood gas sample depending on the ratio of blood to heparin and the buffering ability of the individual blood sample.[1–3]

When using liquid heparin to "heparinize" sampling syringes, all personnel should draw a constant volume of blood to standardize heparin's effect.[2, 4] Too often personnel try to "get by" with a "short sample," which then yields results that do not fit the patient's true medical picture and may lead to incorrect treatment. The PO_2 may be falsely elevated or lowered, the PCO_2 typically will be lowered, and the *p*H will be lowered and acid–base variables altered. The heparin dilution problem has been addressed recently by the introduction of syringes containing dried heparin.[5]

AIR BUBBLES

Another critical source of error in a blood gas sample is air. Air bubbles can grossly alter PO_2 values depending on the amount and size of the bubbles and the PO_2 of the blood sample itself.[6] The smaller the bubbles, the greater the surface area in contact with the blood, and the quicker the variation will occur.

Small, pale, pink, frothy air bubbles are usually caused by an ill-fitting syringe–needle junction; a needle too small for the syringe; a poor-fitting plunger in the syringe; or pulling back too hard on any syringe. Samples with air bubbles must be rejected.

The blood gas sample must be anaerobic. If a sample just drawn contains a small air bubble, the bubble must be expelled within 20 seconds. The syringe should then be made air tight by capping or embedding the needle in a stopper. Bending the needle does not make the syringe air tight and is very dangerous to anyone handling the sample.

COOLING

Preventing or slowing down the metabolic processes to maintain the current state of the specimen is another concern. The most common way is to cool the specimen by plunging the air- and water-tight sample into a container of ice water immediately after drawing the sample, thus cooling the specimen rapidly and slowing the metabolic rate of the white blood cells, which are the major oxygen consumers.[4]

TIME

The specimen must be delivered to the laboratory for analysis within 15 minutes.

CLOTS

Any sample containing clots should be rejected for analysis. Clots tend to form during a difficult puncture when blood does not mix well with heparin or is static in the needle. Clots also form if there is an inadequate supply of heparin in the syringe or if the sample is not mixed immediately after being drawn.

HAZARDS OF ARTERIAL SAMPLING

HEMATOMA

Pressure must be applied to the site immediately after the needle has been withdrawn and maintained for a minimum of 5 minutes. To be certain that the flow of blood through the artery has not stopped, the pulse should be felt through the gauze while pressure is applied. One must be aware that patients on anticoagulant therapy or those with liver disease may bleed longer and may tend to form hematomas or to bleed externally. An arterial puncture site is more likely to bleed than is a venipuncture site because of higher pressure in the arteries.

Usually elastic tissue in the arterial wall tends to cause closure of a hole. With aging and certain disease states, however, the elasticity of tissue decreases and stopping the flow of blood after puncture becomes more difficult. The larger the diameter of the needle, and therefore the hole, the greater the probability of blood leakage.[7] Pressure bandages are not advisable for routine arterial punctures because they may cut off circulation if too tight or allow bleeding if too loose.

ARTERIOSPASM

A transient reflex constriction of the artery in response to pain or other nervous stimuli, arteriospasm can be induced by needle stimulation from a shaky hand. When this occurs, obtaining blood may be impossible, even though the needle is properly located in the lumen of the artery. Arteriospasm may also result in temporarily impaired blood supply to the tissue served by the artery.[7]

THROMBOSIS

A thrombosis (adherent clot) forms if the intima (inner wall) of the vessel is injured. Thrombi tend to form over time when a needle or cannula is left in place; they rarely form as a result of a single arterial puncture. Thrombi may occur both in arteries and in veins; in arteries, they have more serious consequences because not all arteries have collateral vessels to provide adequate alternate blood supply. The safety of a particular site for an arterial puncture primarily depends on the presence of collateral vessels.[7]

SITE SELECTION FOR ARTERIAL PUNCTURE

The process of selecting a site requires some knowledge of anatomy to make a safe decision. The criteria for selection include the collateral blood flow accessibility and the size of the artery and periarterial tissues (fixation of artery and danger of injury to adjacent tissues).[7]

RADIAL ARTERY

The radial artery (Fig. 3-2), small but easily accessible at the wrist in most patients, is the most commonly used site for arterial punctures in clinical situations. The radial artery normally has collateral circulation provided by the ulnar artery, whose presence should be confirmed by Allen's test (*see* p. 69). If the ulnar artery is not present or not functional, the radial site should not be used. The median nerve usually does not lie close enough to the artery to be a hazard. The radial artery can be compressed easily because of the lack of excessive tissue and the presence of bony ligaments. Hematomas are minimal. Thrombosis usually does not occur at this site with single arterial punctures.

The radial artery lies at the wrist medial to the radius and superficial to it when viewing the upturned hand (palm side up). The arm should rest on the bed. The elbow should be extended fully and the wrist extended about 30° with the palm up.

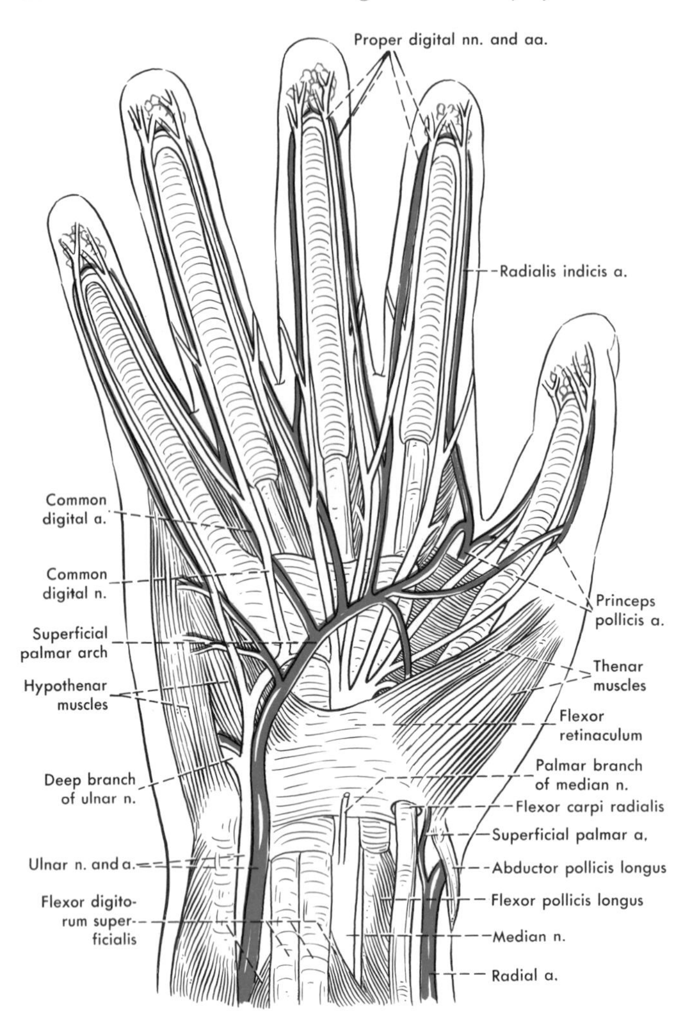

Fig. 3-2. *Anatomy of radial artery region. (Hollinshead WH: Anatomy for Surgeons, Volume 3, 2nd ed, p 492. Hagerstown, Maryland, Harper & Row, 1969)*

BRACHIAL ARTERY

The brachial artery in the arm (Fig. 3-3) usually is the next choice for an arterial puncture. It is larger but more difficult to puncture than any of the superficially located arteries. The brachial artery is located in soft tissue deep between muscles and connective tissues at the anterior side of the antecubital fossa (elbow);[8] its location makes it more difficult to compress effectively, which results in more hematomas during the blood sample procedure. In obese or muscular patients the brachial artery may be impossible to palpate. Good results can be obtained in infants using brachial artery puncture.

The median nerve lies very close to the brachial artery, and, because this

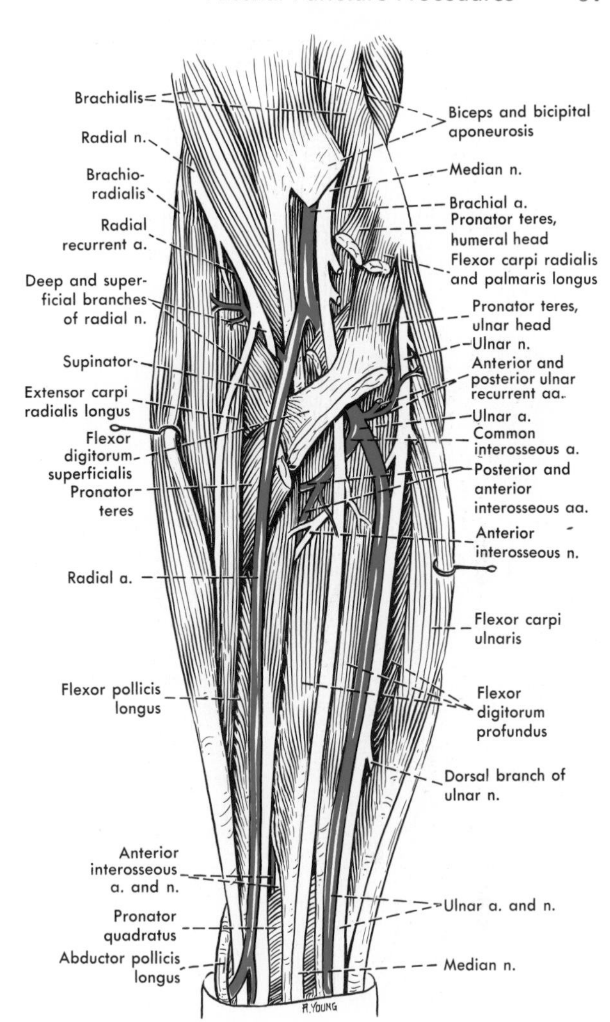

Fig. 3-3. *Anatomy of brachial artery region. (Hollinshead WH: Anatomy for Surgeons, Volume 3, 2nd ed, p 412. Hagerstown, Maryland, Harper & Row, 1969)*

artery is hard to palpate and deeper, it is possible to strike this nerve accidentally with the needle. This area also contains veins that can be punctured accidentally. The phlebotomist must be careful of veins with intravenous (i.v.) fluids running into them; these veins can easily be struck, and the blood will look arterial because of the dilution factor with the i.v. solution. If this occurs, strange results will be obtained on the blood gas measurement.

FEMORAL ARTERY

The femoral artery (Fig. 3-4), superficially located in the groin and easily palpated, usually is the largest artery available but has the most disadvantages. The femoral artery and vein lie in front of the fascia covering the ilieopsoas and

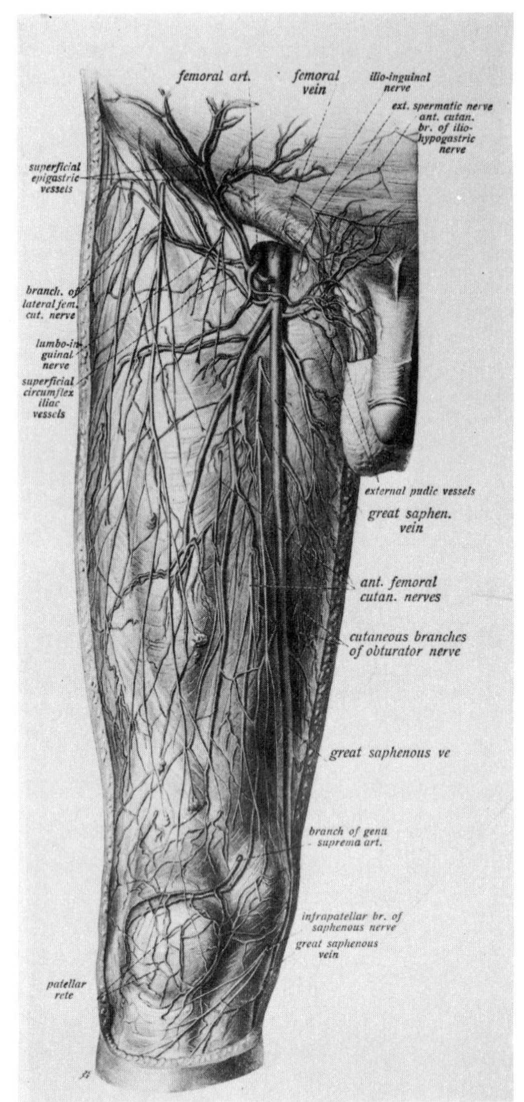

Fig. 3-4. *Anatomy of femoral artery region. (Clemente CD (ed): Anatomy: A Regional Atlas of the Human Body. Munich, Urban & Schwarzenberg, 1975)*

pectineus, with the nerve lying behind. The artery appears midway between the anterior–superior spine and the pubic tubercle and disappears (about 3.5 inches below) where the medial border of the sartorius crosses the lateral border of adductor longus, forming the 'femoral triangle.'"[8]

Although easily punctured, the femoral artery has poor collateral circulation to the leg and a tendency in older patients to build up plaques of cholesterol or calcium on the inner wall of this artery. (These plaques can be dislodged by a penetrating needle.) Other disadvantages include increased risk of infection of the site if not thoroughly cleansed and the presence of pubic hair, which makes aseptic techniques more difficult.[7]

In newborn infants, the hip joint and the femoral vein and nerve lie so close to the artery that possible injury to these structures may contraindicate using this site. In older infants and children, however, puncture of the femoral artery is easy and safe.[7]

Hemorrhage is another possible complication because of the femoral artery's depth from the skin surface, the soft tissue around the artery (with little support for compression of the artery), and the physiologic "empty space" that would allow pooling of blood if bleeding should occur (the size of the artery and the blood pressure therein). A good policy is to apply pressure at this site for a minimum of 10 minutes after the puncture procedure.

Obstruction by thrombi is another area of concern but rarely has been reported because of the large diameter of the femoral artery.[7] Other drawbacks include hitting the femoral nerve and the formation of an arteriovenous fistula.

TEMPORAL ARTERY

The temporal artery, especially in infants, may be as wide as, or wider than, the radial artery and may be punctured easily. One of the two main branches of the temporal artery usually is used.

DORSALIS PEDIS ARTERY

In adults, the dorsalis pedis artery on the top of the foot is an excellent site in special circumstances, such as burns, casts, or multiple injuries to the arms and pelvic region.

EQUIPMENT FOR ARTERIAL PUNCTURE

KITS VERSUS "DO-IT-YOURSELF"

Decisions on equipment will depend on anticipated test volume, the people involved in arterial puncture, the budget, the type of blood gas analyzers, and the criteria used by the physician evaluating the results of the blood gas analysis. If only 10 to 15 arterial punctures per week per facility are anticipated, the use of manufactured kits should be considered. Under any circumstances manufactured kits should be available on cardiopulmonary resuscitation "crash carts" because they are easy for the staff to handle.

Technicians analyzing blood gas should check the particular syringe in any kit for analytical handling. In a life-or-death situation, nothing is more frustrating than not being able to sample from a syringe. A poor-fitting barrel/plunger so loose that it falls out or introduces air into the sample, or a plunger too tight to manipulate for drawing the sample, causes many problems. A phlebotomist wanting particular features in arterial puncture apparatus should make an appropriate selection (Fig. 3-5).

In the larger facility it is more economical to purchase individual components and to prepare the syringe. As much as four to five times the price of a manufactured kit can be saved.

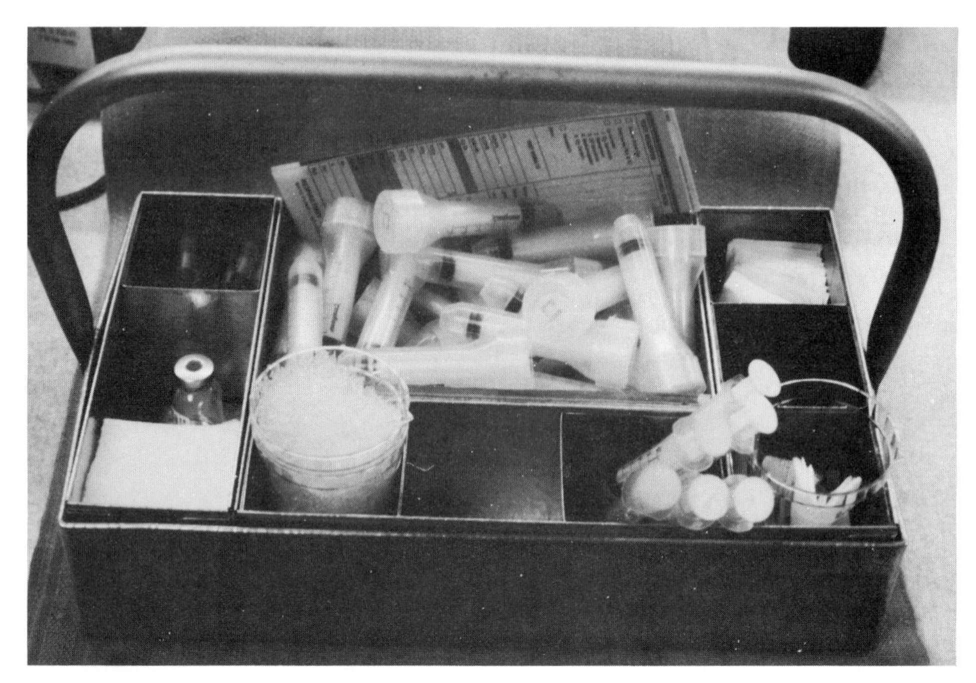

Fig. 3.5. *Example of arterial phlebotomy basket.*

PLASTIC VERSUS GLASS SYRINGES

Most samples obtained for blood gas analysis are not fully saturated with oxygen. Thus, clinically, the oxygen saturation and oxygen content are more important than the PO_2, which is effected by diffusion through plastic and will change over time. The argument for using glass syringes over plastic ones is that oxygen diffuses through the walls of the plastic syringes over 1 to 2 hours.[1, 9, 10] Because time is now becoming a lesser factor, however, this increasingly is not a problem. Technical advances in today's blood gas instrumentation now allow analysis in 1 to 2 minutes, whereas it used to take 15 to 45 minutes to analyze a sample if all went well. Today 90% of samples are totally analyzed within 15 minutes from the time the needle punctures the skin.

Most plastic syringes on the market have excellent barrel and plunger fit and require very little or no aspiration. Drawing arterial blood with a plastic syringe is different than with a glass syringe: Arterial samples can be drawn in plastic syringes without excessive suction that would lower the gas pressure of the sample, and therefore the partial pressure of the individual gases. Proper technique must be taught with each type of syringe and *followed closely.*

A problem with glass syringes is sterilization if they are reused. The major problem, however, is sample handling during analysis: If they are dropped while being mixed, they may break, and if the plunger is too loose, it may fall out if the technician loses his grip for a split second, resulting in altered values due to room air contamination.

Vacuum tubes are not valid for arterial blood gas analysis. The partial pressure of the dissolved gases is altered severely by the suction created by the vacuum in the tube.[11] The filling volume has a definite effect on the partial pressures in the blood sample. When this occurs, there is no consistency in results. Another problem is that an anaerobic system cannot be maintained during analysis of the sample. An appropriate arterial collecting device for arterial blood gas samples should always be used.

Capillary sticks are not recommended for an accurate arterial blood gas analysis. An obvious problem is room air contamination of the sample. There are fewer contradictions for arterial puncture in infants than adults, and there is less blood loss in an arterial puncture.[7, 12, 13]

PROCEDURES FOR ARTERIAL PUNCTURE

EQUIPMENT AND SUPPLIES NEEDED

Antiseptic Solution

Alcohol is the most commonly used antiseptic solution. Iodoform-type antiseptic may be used except that it may contaminate the sample with potassium, contraindicating sodium and potassium analyses. Some persons are allergic to iodine compounds.

Needles

Hypodermic needles, 20, 21, 22, 23, or 25 gauge, 1.5 cm to 3.8 cm (5/8 in to 1½ in) length, may be used depending on the size of the artery and the amount of blood needed. The needle used most commonly is the 22-gauge, 1½-inch needle, a size and length that work well for most adults and for most sites (creating the greatest blood flow and the least injury to the artery). A smaller gauge may be used for infants. The longest needles usually are needed for the femoral and brachial arteries. Quite often a smaller gauge needle facilitates the puncture of tough arteries, which tend to "roll" in older patients. With a 25-gauge needle, blood may not flow into the syringe under its own pressure but has to be aspirated *gently;* is is difficult to obtain more than 1 ml or 2 ml without excessive suction.

Syringes

Syringes or other collection devices should be no larger than the amount of blood needed to avoid excessive dilution of the blood by the anticoagulant solution used to fill the dead space of the syringe and needle. If a sampling probe must be inserted into the syringe, there must be enough depth to aspirate an adequate sample.

Heparin

Sodium or lithium heparin (1000 units/ml) is ideal for blood gas analysis. Oxalate, EDTA, citrate, or other blood preservatives are not acceptable for this procedure. Only enough heparin to fill the dead space of the syringe and the needle is needed and will anticoagulate the amount of blood held by that syringe. A common misconception is that the presence of heparin in the dead space serves to keep air out of the syringe. Because the PO_2 and PCO_2 of heparin

Name: ______________________

Room Number: ______________________

Clinic Number: ______________________

F_{IO_2}: ______________________

Tidal Volume: ______________________

Frequency: ______________________

$P_{cm\,H_2O}$: ______________________

MV c̄ CPP: ______________________

SV c̄ CPP: ______________________

Sample Site: RRA ☐ RBA ☐ RFA ☐

LRA ☐ LBA ☐ LFA ☐

Electrolytes: ______________________

Other: ______________________

Temp: ______________________

Physician: ______________________

PD-1608

17812

TESTS REQUESTED	PATIENT DATA
☒ Arterial Blood Gas	Name: Jack Frost
☐ Electrolytes (NA & K)	OR# 31 Date 10-13-80
☒ Co-oximeter	Clinic No. 2-762-267
1. O_2 Saturation	Sex M
2. Hemoglobin	Physician Anesth.
3. CO Saturation	
4. Methomoglobin	
5. O_2 Content	
☐ Hematocrit	
Other:	

TEST RESULTS

Hb	13.1
%O_2Hb	97.9
%COHb	2.4
%MetHb	0.2
O_2 Ct.	17.8
TEST NO.	454
pH (37°C)	7.464
PCO_2 (37°C)	034.0
PO_2 (37°C)	032.4
BE	+00.6
HCO_3	023.7
CO_2ct	024.7
pH (CORRECTED)	7.538
PCO_2 (CORRECTED)	027.2
PO_2 (CORRECTED)	028.7
TEMP (CORRECTED)	032.0
Na	
K	
Hct	%
P_{50}	mmHg

Temp 32.0 °C/°F
Time 10:50 a.m./p.m.
Sample Site cann
Initial LC

INSPIRED OXYGEN

FIO_2 .40
Other
Tidal Volume .675
Frequency 20
$PcmH_2O$ 25
MVc̄CPP 5
SVc̄CPP
Dead Space —

PATIENT POSITION

Prone

RESPIRATORY INTENSIVE CARE
MAYO CLINIC – ROCHESTER, MINNESOTA

MC 580

Fig. 3-6. *Two examples of data collection forms.*

are presumably the same as that of room air, the dilution effect of heparin on the blood will be the same as that of room air. To minimize and to standardize this solution error, syringes should be filled each time with the same amount of blood.[2, 3]

Capping Devices

The arterial sample must be anaerobic to prevent room air contamination. A Leur tip cap or another suitable capping device must therefore be used. Bending the needle makes a sample dangerous to handle, not air tight. The needle may be embedded in a stopper, if necessary, but this is not recommended.

Gauze

A clean, dry gauze square can be used to pressure the puncture site.

Labels and Forms

The syringe must be labeled in such a way that the identification does not wash off in ice water. Permanent laundry-marking pens work well on the barrel of the butt of the plunger.

Data collection can be recorded on requisition forms or on data pads (Fig. 3-6).

Ice Water

A container with ice water or other coolant should be used to maintain a temperature of 1°C to 5°C and should be large enough to immerse the barrel of the syringe or collection device. This temperature will slow the metabolism of the white blood cells, which continue to consume oxygen.

The blood must, then, be iced until analysis. Arterial samples from leukemic patients with high white blood cell counts should be analyzed within 5 minutes. Samples with anticipated high oxygen tensions, for example, during cardiopulmonary bypass to shunt measurements, should also be analyzed within 5 minutes.

Oxygen Analyzer

An oxygen analyzer is used with patients who are breathing oxygen-enriched gas mixtures in a closed system, such as ventilator or hood (Fig. 3-7).

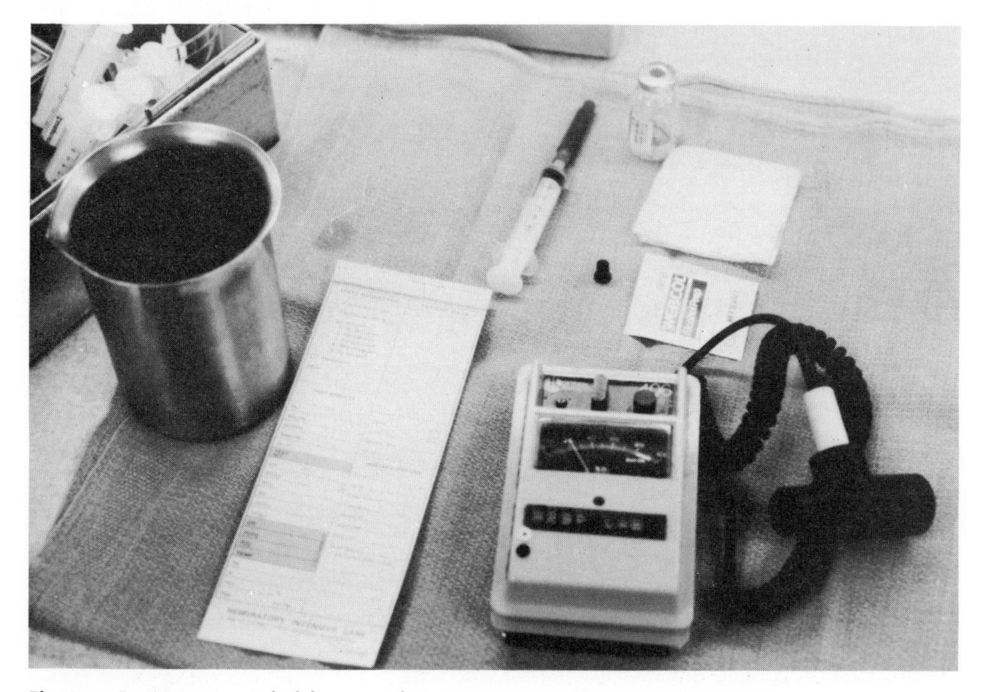

Fig. 3-7. *Equipment needed for arterial puncture.*

PREPARATION OF SYRINGE FOR BLOOD SAMPLING

1. Wash hands thoroughly.
2. Aseptically prepare the top of the heparin vial.
3. Attach a needle to the syringe using aseptic technique.
4. Draw about 0.3 ml of heparin into the sampling syringe.
   - When using a glass syringe, thoroughly wet the entire inside of the barrel of the syringe by moving the plunger up and down.
   - When using a plastic syringe, just move the plunger down a short way to allow complete coating of the rubber end of the plunger.
5. Replace the aspiration needle with the needle selected for the arterial puncture.
6. Holding the syringe upward, expel the air, then, inverting it, expel the excess heparin through the needle, leaving the entire *dead space* of the syringe *and the needle* filled with heparin. Make sure that no air bubbles are left in the syringe or needle. The outside of the needle should not be wet with the anticoagulant solution. Carefully replace the needle in the protective sheath.

PATIENT PREPARATION AND INFORMATION

The patient's temperature, breathing pattern, and the concentration of oxygen in the inspired air (FIO_2) influence the amounts of oxygen and carbon dioxide in the blood. These factors need to be recorded at the time of arterial puncture to allow interpretation of the results. It is also necessary to verify that the patient is at steady state unless an emergency situation exists. Sufficient time should have elapsed between a change in ventilation or FIO_2, 15 to 20 minutes depending on the degree of the patient's ventilatory problems.

The patient's history should be checked for sensitivity to local anesthetic, if he is to be given anticoagulant therapy, if he has bleeding disorders or if he has had surgery or disease in the area adjacent to the contemplated puncture site. These factors should be considered in deciding on the site for the arterial puncture.

1. Obtain the patient's temperature. A policy on temperatures should be established and adhered to.
2. Record if patient is breathing room air, 21% oxygen, or an enriched oxygen mixture. This is reported as FIO_2 fraction of inspired oxygen.
   - If patient is on any oxygen delivery system that allows room air entrainment, such as nasal cannula or T-piece face mask, record only the percentage of oxygen blend.
   - If patient is on a closed system that does not allow room air entrainment, such as a ventilator, hood, or tent, verify the FIO_2 with an oxygen analyzer.
3. If the patient is breathing spontaneously, provide for his physical and mental comfort by quiet and reassuring talk and a comfortable position.[14] The patient should be relaxed, lying comfortably in bed for at least 10 minutes. Meanwhile no changes should have occurred that would effect the blood gases.

The following would cause significant changes: change in respiratory rate, tidal volume, and FIO_2. Also effecting blood gases would be tracheal suctioning, intermittent positive pressure breathing treatment, and chest physical therapy. Blood gas values will be altered temporarily by hyperventilation owing to anxiety or pain, breath-holding, or crying.

PERFORMANCE OF THE PUNCTURE

The puncture site must be selected carefully using the criteria previously discussed. Special arm bands may warn of problems that will affect the choice of sites, including such things as mastectomies, femoral grafts, vascular diseases, or fistulas.

Radial Artery

Before this site is selected, the presence of adequate collateral circulation through the ulnar artery should be established by Allen's test. In this procedure, the patient's hand is closed tight in a fist. The phlebotomist should apply pressure and obstruct both the radial and ulnar arteries (Fig. 3-8). The patient should then rapidly open and close his hand (or have it done for him) until the palm and fingers are blanched. He should then leave his hand open. The phlebotomist releases only the ulnar pressure and observes the whole hand, which should become flushed within 15 seconds as the blood from the ulnar artery refills the empty capillary bed. If the ulnar artery does not supply the entire hand adequately—a negative Allen's test—the radial artery should not be used as a puncture site. If the Allen's test is positive, the site may be used.[7, 15] The patient's arm should be abducted with the palm facing up and the wrist extended to about 30° to stretch and fix the soft tissues over the firm ligaments and bone. If needed, a rolled bath towel may be placed under the back side of the patient's wrist. If assistance is available, have that person hold the arm securely under the elbow and across the palm of the hand after it has been positioned.

A finger should be placed carefully over the artery just proximal to the wrist skin crease to palpate for size, direction, and depth of the artery. The phlebotomist should be in a comfortable position before attempting the puncture—even sitting down if it affords better body mechanics.

The puncture site must be prepared aseptically. After cleansing, it must remain untouched by fingers. After preparation of the syringe as described above, the phlebotomist must make sure that *no air is drawn into the syringe* again. Holding the syringe in one hand like a dart, the phlebotomist then places a finger of the other hand over the artery at the exact point where the needle will enter the artery (not the skin).

The skin is punctured about 5 mm to 10 mm distal to the finger directly over the artery using a 30° angle of insertion with the bevel of the needle up (Fig. 3-9). After the needle penetrates the skin, it should be advanced slowly, aiming for the artery just beneath the finger. As the syringe is advanced, the flash-back chamber should be observed for blood as the artery is entered. If a plastic syringe or a needle smaller than 23 gauge is used or if the patient is hypotensive, it may be necessary to pull *very gently* and *slowly* on the plunger

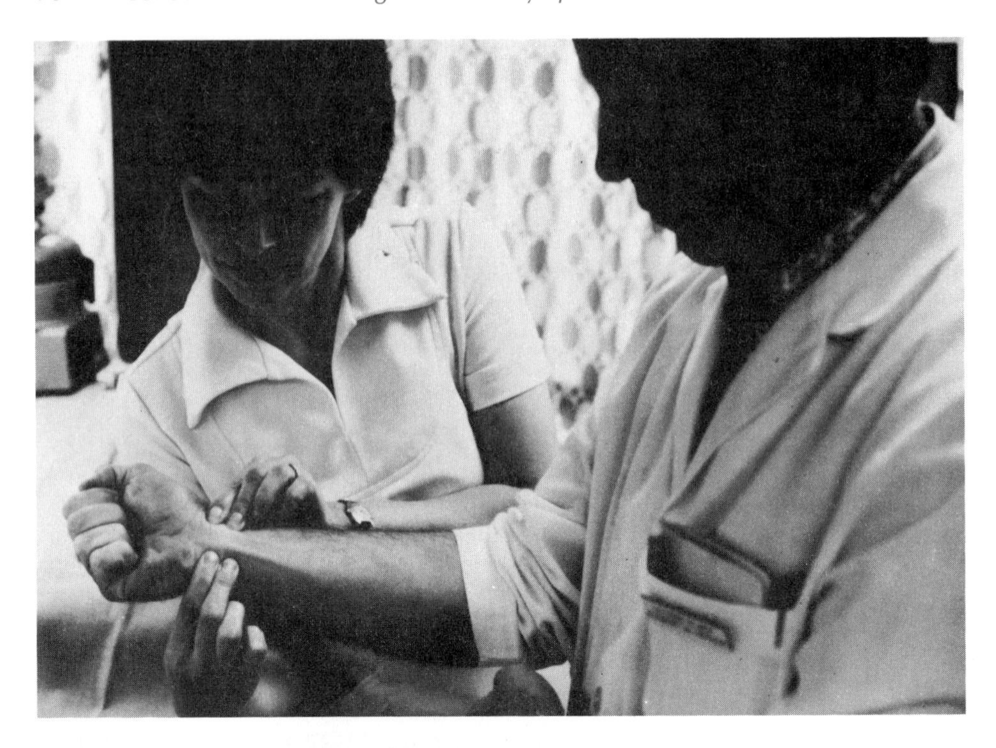

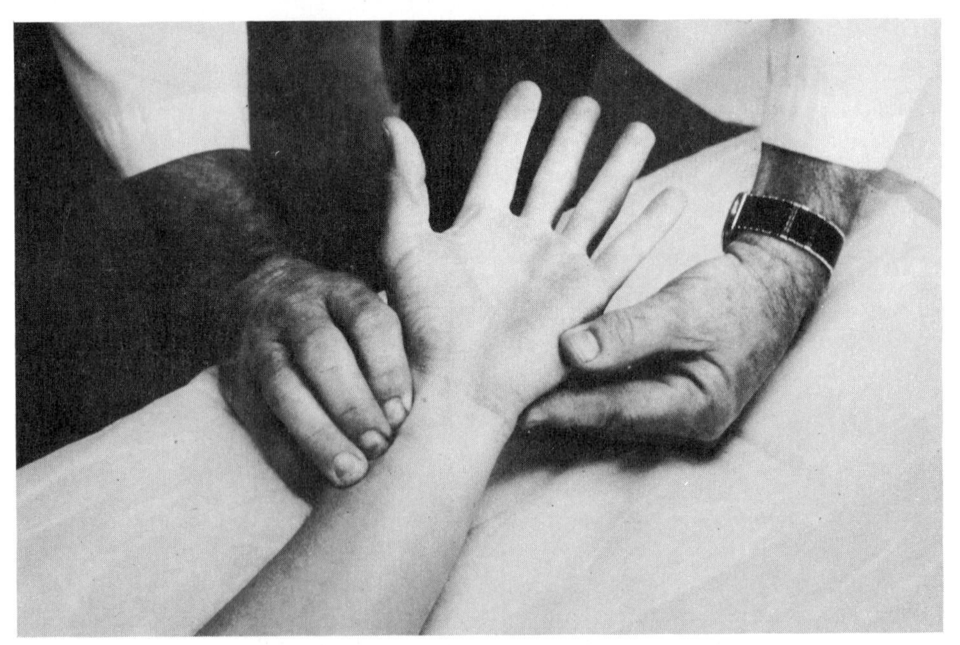

Fig. 3-8. *Phlebotomist occluding radial and ulnar arteries while patient clinches fist* (top). *Radial artery occluded while ulnar artery pressure is released and patient's hand is relaxed to observe flow of arterial blood into palm* (bottom).

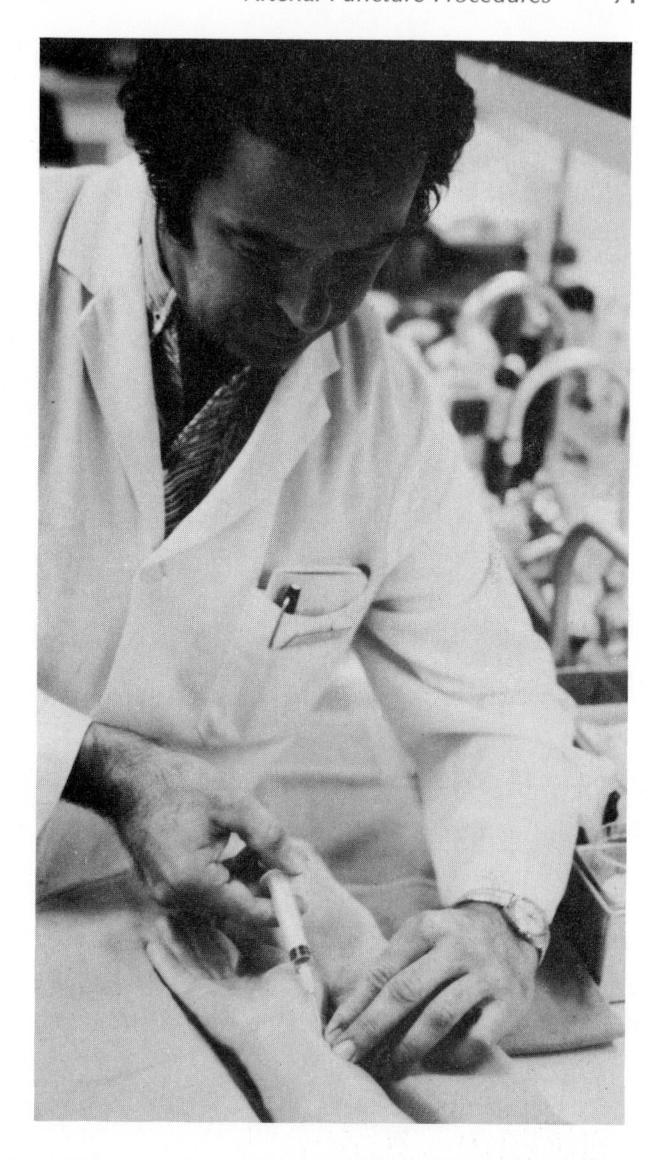

Fig. 3-9. *Correct angle of radial artery puncture.*

for blood to flow into the syringe. If a glass syringe is used and the patient is not hypotensive, the pressure will push the plunger back spontaneously. Pressure should be applied on the end of the plunger to prevent it from being pushed out. Once the artery is entered, the hand that palpated the artery may be used to help manipulate the syringe.

After obtaining the necessary amount of blood, the phlebotomist should quickly withdraw the needle and syringe, simultaneously placing a dry gauze sponge over the puncture site and the needle in its protective sheath. Immediately and continually, firm pressure should be applied at the puncture site for a minimum of 5 minutes but the artery must not be occluded. A pulse should be felt through the gauze. If the patient is receiving anticoagulant or has a prolonged clotting time, the pressure on the site should be held longer. Two minutes after relieving pressure, one should inspect the puncture site again to ascertain that no hematoma is developing. Pressure dressings are not acceptable on an arterial puncture site. If the phlebotomist cannot stay to hold the site, the nurse should hold it, with explicit instructions on how long and what to look for. If bleeding cannot be stopped within a reasonable time, the physician in charge should be notified. While pressure is being applied on the puncture site, the syringe should be checked for air bubbles. If present, they should be dislodged by holding the syringe with the needle up and carefully pushing any air out of the syringe.

The needle is removed, and the syringe is capped with a leur tip cap or the needle is embedded in a stopper to make the syringe air tight and impervious to water. Bending the needle is TOTALLY UNACCEPTABLE! The sample is labeled in such a way that the label is waterproof (permanent laundry marker pens work well). The blood sample is then immersed in an ice water bath immediately (Fig. 3-10).

Another method for arterial puncture is the use of a winged infusion set with a small needle. .The cannula is inserted into the artery with the above technique without an attached syringe. When blood enters the plastic catheter by pulsatile flow, it is allowed to fill the catheter and the hub of the adaptor before the syringe is attached. With this technique, several samples may be obtained for different determinations without dislodging the needle in the artery and without air bubbles.[7]

Brachial Artery

The brachial artery, deeper than the radial artery, lies beneath the basilic in the medial or ulnar and anterior aspect of the antecubital fossa.[8] The position of the patient and the preparation of the syringe are the same as described above. Caution and skill in performing the puncture are needed to avoid hitting the median nerve, which lies very close to the brachial artery.

The patient's arm is fully extended to locate the artery and the wrist rotated until the maximum pulse is palpated with the index finger just above the skin crease in the anticubital fossa. The artery should be palpated proximally with the middle finger for 2 cm to 3 cm for size, direction, and depth of the artery.

The puncture site is next cleansed. While he places his fingers 2 cm to 3 cm apart along the pulsations of the artery, he punctures the skin about 5 mm to 10 mm distal to his index finger using a 45° angle of insertion with the bevel. He directs the needle along a line connecting the two fingers, avoiding visible or palpable veins. The brachial artery lies deep in the tissues in most persons, especially in those who are obese; it does not run parallel to the bone. After withdrawing the needle, the phlebotomist should compress the artery for a minimum of 5 minutes or until the bleeding has stopped.

Femoral Artery

The femoral artery and vein lie just below the inguinal ligament in front of the fascia covering the iliopsoas and pectineus with the nerve lying behind. The artery appears midway between the anterior–superior spine and the pubic

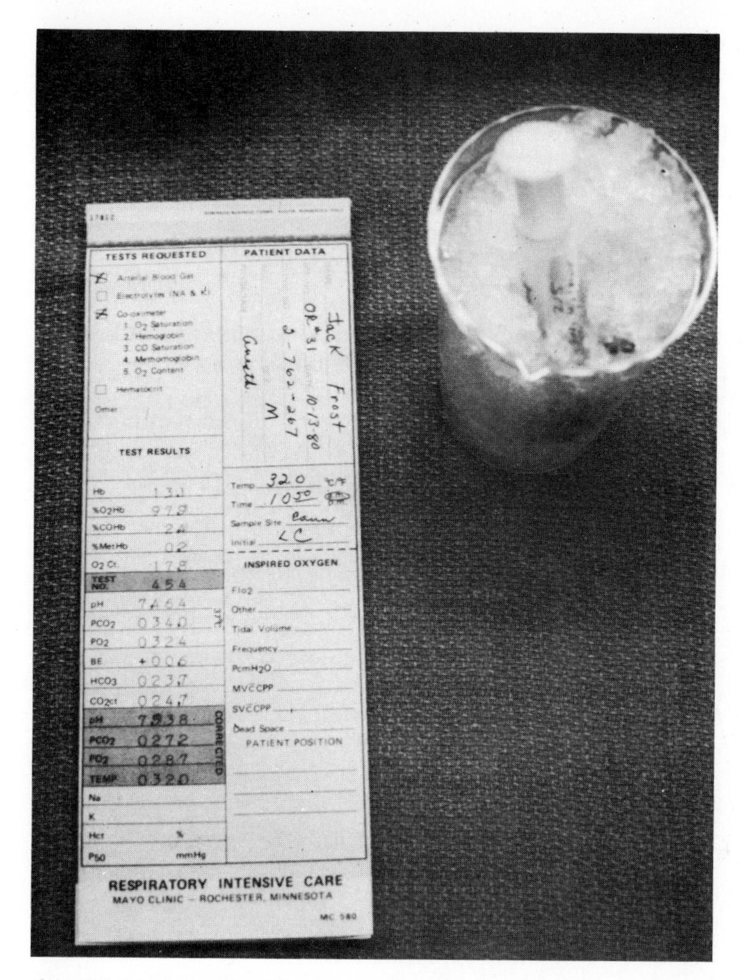

Fig. 3-10. *Iced sample with form.*

tubercle and disappears (about 3½ inches long) where the medial border of the sartorius crosses the lateral border of adductor longus, forming the femoral triangle.[8]

The patient should lie flat with both legs extended. The phlebotomist then palpates the pulsating vessel with two fingers. When unable to palpate a pulse on an adult, he should enter the skin two fingers distal to the inguinal ligament at the midline between the anterior–superior iliac spine and the symphysis publis. The femoral artery site should *not* be used if the patient had a femoral artery graft. A graft should be suspected if a scar is noted in this area and must be documented by reviewing history.

The puncture site must be thoroughly cleansed because the groin area typically is heavily contaminated. The artery is penetrated between the fingertips (spread 2–3 cm apart) at a 45° to 90° angle. The needle bevel faces the flow of blood.

After the required sample is obtained, the needle is withdrawn. Firm pressure should be applied on the artery for 10 minutes or until the bleeding has stopped. The femoral artery is deeper and has more blood pressure than the other arteries and requires more pressure and time for clotting after puncture. The femoral artery site can lead to the most serious complications, and thus extreme caution should be used.

Sampling from Arterial (or Venous) Catheters and Cannulae

Needle cannulae or catheters have an anticoagulant flush solution infused continuously through them to prevent clotting in the cannula or catheter. This flush solution must be interrupted to obtain a blood sample.

Air must not be introduced into the system and all connections should be secure. Before sampling, all the "dead space" contents of the catheter and connectors must be removed completely.

All other criteria for arterial sampling must be followed.

With a sterile syringe the "dead space" contents are aspirated using a stopcock inserted between the cannula or catheter and the infusion line. The volume of aspirated liquid depends on the inner diameter length and geometry of the catheter or cannula and connectors. A uniform diameter and smooth shape facilitate removal of the flushing solution. Once blood reaches the syringe, an additional small amount should be withdrawn to prevent any dilution of the blood. The second heparinized sample syringe is then inserted into the stopcock, and a sample is slowly aspirated into the syringe. The syringe is removed, capped, and iced.

The sampling port is cleaned aseptically of all traces of blood, usually by manipulating the stopcock to allow flush solution to purge the port. The waste is caught with sterile gauze or a sterile receptacle. The stopcock is then manipulated to reestablish the flow of flush solution, and an additional small amount of the flush solution is infused into the cannula or catheter to ensure its being cleared of blood and that the normal infusion rate has resumed.

Local Anesthesia

Local anesthesia should not be used indiscriminately. On rare occasions, local anesthesia may be needed at the site of an arterial puncture for an uncooperative patient whose steady-state situation has been destroyed. (It then acts to separate the pain.) Experience has shown that the introduction of the anesthetic under the skin is as painful as the actual arterial puncture. A skilled phlebotomist can obtain the arterial sample quickly and deftly so as to cause little discomfort to the patient. Local anesthesia need not be used in already cannulated patients, artifically ventilated patients, unconscious patients, and newborn infants. Local anesthesia should not be used in patients who will need multiple arterial punctures over time because the number of times the skin has been punctured would double and lead to complications and ulceration.

Local anesthesia may be used in certain situations to avoid struggling, vigorous crying, breath-holding, or hyperventilation during the arterial sampling. In this situation, procaine 1% or xylocaine 1% may be used.

The contents of a vial of procaine or xylocaine are drawn into a 3-ml syringe using a 25-gauge needle. Air bubbles are removed, and the needle is put in its protective sheath. The chart should be checked and the patient asked whether he is allergic to the anesthetic. The sampling site is then chosen and the skin cleansed. Where the skin puncture will be made, the needle is inserted just under the skin, avoiding any superficial blood vessels. The phlebotomist pulls back slightly on the plunger of the syringe to verify that he is not in a blood vessel. He raises a small wheal by injecting about 0.5 ml of procaine or xylocaine. After waiting 1 to 2 minutes for the anesthetic to take effect, he should then proceed as usual with the arterial puncture.

The importance of establishing procedures for arterial puncture cannot be emphasized enough. Arterial blood gas values can be affected more by sample handling than any other factor: More erroneous results have been reported from poor sampling technique than from analytical error. Discipline is paramount, along with having thoroughly trained personnel. Training should be standardized throughout the institution as a form of quality control. Only then can arterial puncture become a reliable clinical aid.

REFERENCES

1. Bageant RA: Variations in arterial blood gas measurements due to sampling techniques. Respir Care 20:565–570, 1975
2. Cissik JH, Salustro J, Patton OL, Louden JA: The effects of sodium heparin on arterial blood gas analysis. Cardiovasc Pulmonary Technol J 9:17–20, 35, 1977
3. Gast LR, Scacci R, Miller WF: The effect of heparin dilution on hemoglobin measurement from arterial blood samples. Respir Care 23:149–154, 1978
4. Van Kessel AL: The blood gas laboratory, an update: 1979. Lab Med 10:419–429, 1979
5. Gauer P, Friedman J, Imrey P: Effects of syringe and filling volume on analysis of blood pH, oxygen tension, and carbon dioxide tension. Respir Care 25:558–563, 1980
6. Mueller RG, Lang GE, Beam JM: Bubbles in samples for blood gas determinations: A potential source of error. Am J Clin Pathol 65:242–249, 1976
7. National Committee for Clinical Laboratory Standards (NCCLS): Standard for the Percutaneous Collection of Arterial blood for Laboratory Analysis. Villanova, Pennsylvania, NCCLS, 1980
8. Ellis H: Anatomy for anesthetists. Anaesthesia 16:235–240, 1961
9. Evers W, Racz GB, Levy AA: A comparative study of plastic (polypropylene) and glass syringes in blood gas analysis. Anesth Analg (Cleve) 51:92–97, 1972
10. Scott PV, Horton JN, Mapleson WW: Leakage of oxygen from blood and water samples stored in plastic and gas syringes. Br Med J 3:512–516, 1971
11. Gambino SR: Heparinized vacuum tubes for determination of plasma pH, plasma CO_2 content, and blood oxygen saturation. Tech Bull Registry Med Technol 29(8):123–131, 1959
12. Kisling JA, Schreiner RL: Techniques of obtaining arterial blood from newborn infants. Respir Care 22:513–518, 1977
13. Folger GM, Kouri P, Sabbah HN: Arterialized capillary blood sampling in the neonate: A reappraisal. Heart Lung. 9:521–526, 1980
14. Thompson C: Blood drawing and your patient's emotions. Medical Laboratory Observer 9:51–54, 1977
15. Greenhow DE: Incorrect performance of Allen's test: Ulnar-artery flow erroneously presumed inadequate. Anesthesiology 37:356, 1972

SUGGESTED READING

1. Bouhortos J, Morris T: Femoral artery complications after diagnosis procedures. Br Med J 3:396, 1973
2. Felkner D: A protocol for teaching and maintaining arterial puncture skills among respiratory therapists. Respir Care 18:700, 1973
3. Lindesmith LA et al: Arterial puncture by inhalation therapy personnel. Chest 61:83, 1972
4. Macon WL, Futrell JW: Median nerve neuropathy after percutaneous puncture of the brachial artery in patients receiving anticoagulants. N Engl J Med 288:1396, 1973
5. Matthews JI et al: Embolization complicating radial artery puncture Ann Intern Med 75:87, 1971
6. Petty TL et al: The simplicity and safety of arterial puncture. JAMA 195:693, 1966
7. Sackner MA et al: Arterial punctures by nurses. Chest 59:97, 1971

4

Blood Collection Procedures

Blood Cultures

M. Marsha Hall

Blood cultures, collected whenever the clinician has reason to suspect clinically significant bacteremia, are one of the most important cultures performed in the Clinical Microbiology Laboratory. Blood cultures help to indicate the severity and extent of spread of an infection. They provide for the identification and antimicrobial susceptibility of the etiological agent causing a severe or life-threatening disease. Therefore, the technique and procedure used in the collection and processing of these specimens are important for proper patient care.

SKIN ANTISEPSIS

The most important procedure performed during the collection process is proper skin antisepsis. Although variable from person to person, many bacteria, both gram positive and gram negative, are present on the skin.[1] Gram-negative organisms are less common inhabitants on normal, healthy skin but are not uncommon on the skin of hospitalized patients or hospital personnel.[2] Therefore, there is a high risk of blood culture contamination from the skin of the patients or, for that matter, from the skin of the phlebotomist collecting the culture. The significance of these organisms, although usually nonpathogenic in nature, may be difficult to establish when they are isolated from a blood culture because of their role in causing endocarditis and because of implanted prosthetic material infections. The laboratory must report all microorganisms isolated from the blood cultures. The physician must interpret the report and decide whether the isolate is clinically significant or whether the isolate is merely a contaminant. If the isolate is interpreted as clinically significant, a designated treatment protocol is indicated. The patient could be committed to additional hospitalization for treatment at considerable expense and some risk because of possible adverse toxic effects due to antibiotics. The role the phlebotomist plays in preparing the site for venipuncture is one of the most important involved in blood cultures. The phlebotomist, by his expertise or lack of it, either contributes to the patient's welfare or possibly causes misleading information to be reported.

EFFECTIVE AGENTS FOR SKIN PREPARATION

The need for careful and proper skin antisepsis cannot be stressed enough to reduce the incidence of contaminated blood cultures. The most effective agents are the following.

1. One percent to 2% tincture of iodine and 70% ethyl or isopropyl alcohol. First, the alcohol is used to clean the skin, and next the iodine is applied in a concentric fashion over the venipuncture site.
2. Iodophors may be used instead of iodine because they are relatively less allergenic and cause less sensitivity or irritation. A commercially available packet of three swabs, soaked in iodophor, can be used and serves multiple purposes: It is effective, convenient, nonallergenic for patients who have iodine skin hypersensitivity, and helps the phlebotomist follow a certain routine.
3. Seventy percent to 95% alcohol. One problem with alcohol is that too frequently large numbers of cotton sponges are stored in a jar containing alcohol in which the alcohol is replenished only when necessary. Evaporation of the alcohol, along with the contamination contributed by fingers and air, substantially reduces the activity and efficacy of the antiseptic agent. To avoid such problems, cotton sponges or pads, impregnated with alcohol and wrapped individually in air tight foil packets, should be used.

PREPARATION OF SITE

"Instant" antisepsis does not occur regardless of which antiseptic agent is used. These agents require at least 1 to 2 minutes before they exert any significant activity against most skin bacteria. It is therefore important to use the first swab with antiseptic agent to scrub the venipuncture site for 1 to 2 minutes. The second swab should be used concentrically on the site. By following this procedure, the contamination rate of blood cultures from 240 adults without infection has been found to be 2.1%, an acceptable level.[3] Because some bacteria live in the skin itself, it is impossible to eliminate contamination altogether. Once the venipuncture site has been prepared aseptically, it should never be touched unless the fingers used for palpation have also been disinfected (in the same manner as the venipuncture site).

PREPARATION OF CONTAINER

A swab with antiseptic agent is used to clean off the rubber stopper or diaphragm top of the blood culture bottle or collection container. The rubber stoppers or diaphragm tops of culture bottles or collection containers are potentially contaminated and must be prepared aseptically before injecting the blood.

SPECIMEN COLLECTION

The following methods are used for specimen collection.

1. Blood can be drawn with a sterile needle and syringe.
2. Blood can be collected with a transfer set consisting of sterile tubing with a needle at either end.
3. Blood can be collected in evacuated blood culture bottles that fit into tube holders used for routine venipunctures.

The transfer set is preferable because it may provide less chance of contamination; however, the syringe is more convenient when blood must be collected for other purposes at the same time. The use of evacuated blood culture bottles is not generally recommended because of the risks of contamination from nonsterile tube holders and of backflow of the culture medium. The blood may be inoculated directly into the bottles of culture medium at the patient's bedside. In some hospitals, the blood is collected in a sterile evacuated blood collection tube that contains anticoagulant and is transported to the laboratory for distribution into culture media. Such an approach may actually reduce the contamination rate of blood cultures at institutions where blood is collected by persons who do not use proper skin antiseptic techniques.

SITE

The site or source of blood collection influences the contamination rate of blood cultures. Cultures of blood from the umbilical or femoral vein are more likely to be contaminated than are those of blood from the antecubital vein. Because indwelling intravascular catheters become colonized with bacteria when left in place for longer than 48 hours, cultures of blood taken from such catheters are more likely to become contaminated than are those of blood collected by percutaneous venipuncture. For future reference, it is important for the venipuncturist to note on the culture request form when blood is obtained for culture from the femoral or umbilical vein or from an intravascular catheter. Such information is essential for proper interpretation of a positive culture.

PROCEDURE

1. Identify patient.
2. Assemble and prepare equipment. Remove screw cap from each culture bottle, if present, and leave caps open-side-up next to bottles.
3. Apply tourniquet while selecting venipuncture site, then loosen.
4. Prepare site. Open povidone-iodine pack. Remove one and use it to clean the venipuncture site by scrubbing a 3-inch to 4-inch square area for 2 minutes. Discard swab. Remove second swab from package. Before using, squeeze swab against foil sides of package to remove excess solution. Cleanse the site again, beginning at the center of the site and scrubbing in a circular motion outward to a diameter of 3 inches or 4 inches for 30 seconds. Let dry. Do not touch prepared site with unsterile fingers. If it is necessary to palpate the site before venipuncture, cleanse finger with povidone-iodine in same manner as site preparation and allow to dry.
5. Cleanse culture bottle tops. Use third swab in package for scrubbing rubber

tops of culture bottles after removing excess solution from swab by squeezing against sides of foil package.
6. Reapply tourniquet and draw volume of blood needed from patient using a sterile needle and syringe and using acceptable procedure.
7. Replace the used needle on syringe with new sterile needle, making sure not to contaminate the outside of the new needle.
8. Inoculate culture bottles by carefully adding correct amount of blood to each bottle. Make sure not to add air to these cultures. Replace screw cap on those bottles that have a cap.
9. Attach label with necessary information including name, identifying number, date and time of collection, and the venipuncturist's initials. Do not cover media area.

The different types of blood cultures drawn include general cultures, such as those for anaerobes, *Brucella,* fungi and yeast, *Actinomyces, Leptospira,* and *Neisseria.*

Institutional policy will vary as to the containers used for blood cultures and the volume of blood needed for inoculation. At our institution we have found it essential that no bottle be inoculated with more than 10 ml of blood per 100 ml of medium. If a larger volume ratio is used, the blood will frequently clot, seriously impairing the recovery of the bacteria.

The volume of inoculate is an important factor because the order of magnitude of bacteremia has been shown to be quite small in most cases for adults. As few as 1 to 30 organisms per milliliter of blood occur in most types of endocarditis. This may differ somewhat in infants when the order of magnitude of bacteremia is generally higher than that seen in adults. In most cases, too, bacteremias are intermittent so that a single set of blood cultures can be fairly misleading. At the Mayo Clinic, the following guidelines for volumes of blood for culture have been established.

Infants and small children up to 2 years
1 ml into one bottle
Children 2 to 5 years
1 ml into each of three bottles
Children 6 to 10 years
2 ml into each of three bottles
Children 11 to 15 years
4 ml into each of three bottles
Children 16 and older
10 ml into each of three bottles
Adults 10 ml into each of three bottles

When more than one culture is ordered but no times for collection have been designated, it is our practice at the Mayo Clinic to collect them at least 1 hour apart. Occasionally it is necessary to collect more than one culture immediately if a patient is to begin antimicrobial therapy, for instance. In such case, each culture must be collected by separate venipunctures. If a skin contaminant is inoculated into one culture, a second venipuncture will eliminate the situation where several bottles are positive with the same organisms. Such results are misleading to the attending physician.

CONTAMINATION OF CULTURES

In many hospitals the collection of blood for culture is performed by the physician. Unfortunately, they are often unable to devote the time and attention to the correct procedure. Contamination rates have been shown to be significantly lower in blood cultures collected by trained venipuncturists than in those collected by house officers. Regardless of the specific reasons for the differences in contamination of blood cultures, there is little question about the value of an experienced phlebotomist in minimizing the work and confusion generated by contamination of cultures. Regardless of how experienced or proficient a phlebotomist may be, there are times when the vein is missed when doing a venipuncture and a second venipuncture is required. When this happens, it is important always to use a new needle or transfer set and repeat the entire procedure. Aseptic technique is of paramount importance.

CONTAMINATION RATES

The laboratory must monitor the rate of contamination that occurs in the blood cultures (Fig. 4-1). Even under optimal conditions, a contamination rate of 2%

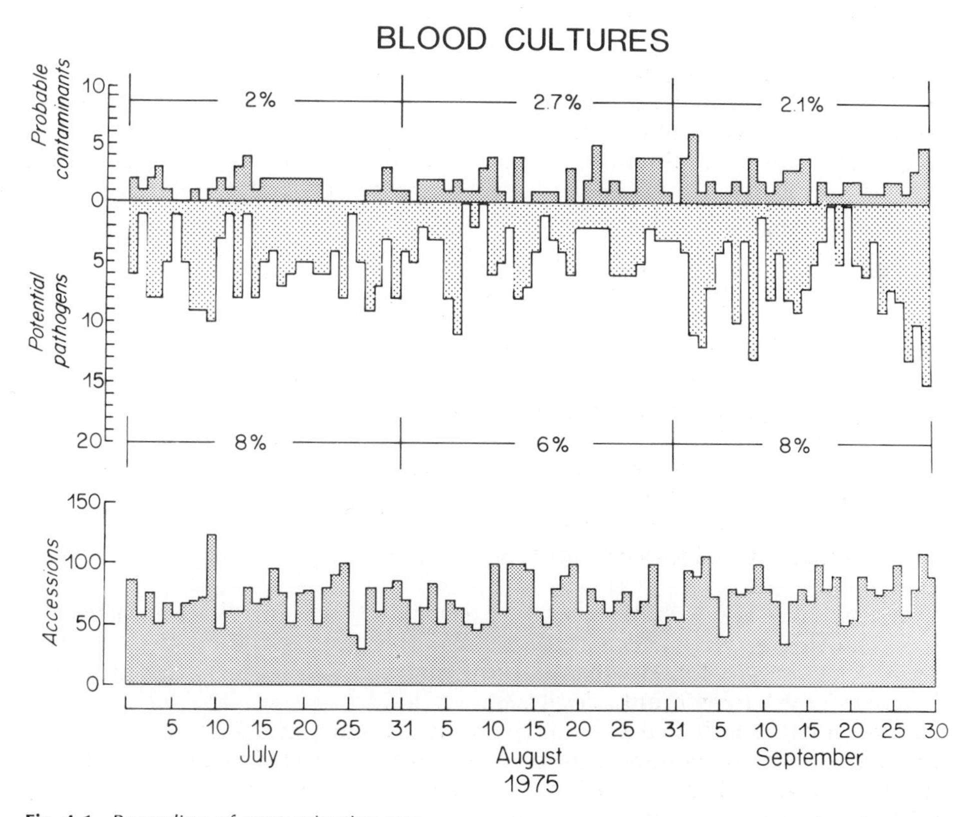

Fig. 4-1. *Recording of contamination rate.*

to 3% can be expected. Presumed contaminants and presumed pathogens can be recorded daily. From the total number of blood cultures performed each month, the percentage of contamination can be determined. At the Mayo Clinic, the monthly contamination rates vary between 0.5% and 2.0%.[4]

If there is a significant increase in contamination, one should look for the source of the serious contamination problem. Obviously there are sources of contamination in blood cultures other than those related to blood collection, such as improper bottle preparation and laboratory processing techniques. The importance of skin as one source cannot be overlooded, however, and does in fact contribute the greatest risks of contamination.

REFERENCES

1. Evans CA, Stevens RJ: Differential quantitation of surface and subsurface bacteria of normal skin by the combined use of the cotton swab and the scrub methods. J Clin Microbiol 3:576, 1976
2. Noble WC, Somerville DA: Microbiology of Human Skin. Philadelphia, WB Saunders, 1974
3. Wilson WR, Van Scoy RE, Washington JA II: Incidence of bacteremia in adults with infections. J Clin Microbiol 2:94, 1975
4. Washington JA II: The Detection of Septicemia. Cleveland, CRC Press, 1978

Preparation of the Peripheral Blood Film

Robert V. Pierre
Mary Morris

Examination of a peripheral blood smear for red blood cell morphology and number, appearance of platelets, and performance of a leukocyte differential count is fundamental for evaluating patients with hematologic disorders. In addition, it frequently is routinely used in the health evaluation examination in many medical practices. The correct interpretation of changes in any of the blood components on a peripheral blood smear depends on the quality of the blood smear and its staining.

The methods of preparation of peripheral blood smears and the advantages and disadvantages of each method, the common errors and artifacts of these methods, and the potential effects on the interpretation of the peripheral blood smear forms the basis of this discussion.

Peripheral blood smears may be prepared from venous blood samples or capillary blood samples. The differences in leukocyte and platelet concentration are slightly different between capillary and venous samples. The differences are not, however, of practical clinical significance to the interpretation of a peripheral blood smear if the samples are collected properly. The preferred method of collection of samples for preparation of peripheral blood films is the capillary sample, with avoidance of the use of anticoagulants. Venous blood samples are acceptable if the blood smears are prepared quickly. A drop of nonanticoagulated blood from the needle tip may be used. If venous blood has been placed in anticoagulant, the smears should be made within 1 hour of collection. Ethylenediamine tetra-acetic acid (EDTA) is the anticoagulant of choice. Samples collected without anticoagulant are preferred because anticoagulants introduce a number of artifacts that may seriously hamper one's ability to interpret properly changes of clinical significance in all blood lines.

ANTICOAGULANT EFFECTS

Ethylenediamine tetra-acetic acid is the anticoagulant of choice for collection of hematologic specimens for cell counting and preparation of peripheral blood smears. Heparin may be used but causes problems with the staining of blood smears because of its strong affinity for basic dyes. The adverse effects of EDTA on preparation of blood smears are affected by length of exposure of the sample to the anticoagulant and the concentration of the EDTA.

EFFECTS ON PLATELETS

Platelets tend to show aggregation on blood smears prepared from capillary skin punctures. This is a normal phenomenon. The lack of platelet clumping on such a sample might indicate defective platelet function and provide a valuable clue to the recognition of platelet disorders such as Glanzmann's disease. Normal platelets have a relative narrow size range and degree of granulation on capillary blood smears. Certain qualitative platelet disorders are characterized by abnormally large platelets, for example, the May–Hegglin anomaly and the Bernard–Soulier (giant platelet) syndrome. Platelets in EDTA anticoagulated blood may show swelling and appear large on blood smear. Anticoagulant action also prevents normal aggregation of platelets. Blood smears prepared from anticoagulated blood therefore cannot be used to recognize disorders in which there is impaired platelet aggregation or abnormal platelet size. On the other hand, EDTA itself will, on rare occasions, cause aggregation of platelets and leukocytes, so that large clumps of platelets or leukocytes may be seen on the blood smear. This clumping takes place in the EDTA blood vial before the blood smear is made; therefore platelet counts and white blood cell counts on such samples by automated instruments will be falsely low. Fortunately, EDTA-induced platelet and leukocyte clumping is a relatively rare event, and instruments such as the Coulter S Plus and Ortho ELT-8 will flag these specimens in most cases.

In one situation the anticoagulated sample is superior for leukocyte differential counts. Image-processing instruments such as the Geometric Data Corporation-Hematrak estimate the adequacy of platelet numbers from blood smears but underestimate the platelet numbers from fingerstick smears because of the normal platelet clumping. The estimate is more accurate on EDTA anticoagulated prepared smears because the platelets are isolated and not clumped.

EFFECTS ON RED BLOOD CELLS

The size, shape, and color of red blood cells on a peripheral blood smear are important findings and may provide diagnostic information or provide significant clues as to the cause of anemia in a patient. Blood smears prepared from EDTA anticoagulated vials may show artifacts secondary to the anticoagulant. These artifacts are minor if the proper proportion of blood and anticoagulant is used and the smear is prepared within 1 hour after collection. Inadequate anticoagulant will lead to clotting of the sample, making it unsuitable for preparation of blood smears. Excessive amounts of EDTA will create artifacts that make red blood cell morphology difficult to interpret.

The artifacts of red blood cell morphology seen in blood smears prepared from EDTA anticoagulated blood include the following.

1. Spiculated red blood cells
2. Target cells
3. Rouleaux
4. Spherocytes
5. Stomatocytes
6. Punched-out red blood cells

In general, the longer that blood is exposed to EDTA in an evacuated tube, the more marked will be the artifactual changes. Smears made within 1 hour from tubes stored at room temperature are acceptable in most instances. But because the appearance of spiculated red blood cells, target cells, rouleaux, and spherocytes may be important diagnostic clues in the evaluation of patients with hematologic disease and it is impossible to distinguish the EDTA artifactual changes from the disease-produced changes, the opportunity to detect and to use these changes in recognition of disease or differential diagnosis is lost.

Spiculated red blood cells are cells with spicules on their surface. The term *burr cells* has been widely used in the past but is a poor term because it has denoted various morphologic red cell forms that have differential mechanisms of formation and different significance. Bessis has suggested the following terminology.[1]

Echinocytes red cells with 10 to 30 spicules distributed evenly over the surface. Spicules are generally short and fine with a sharp point.

Acanthocytes red cells that are small and spheroidal and therefore appear dense. Spicules number 2 to 20 and are distributed irregularly over the surface of the cell. Spicules are longer and thicker and may end in blunt ends or bulblike ends.

Dacrocytes red cells with a single large spicule that deforms the red cell into a tear-drop shape.

Drepanocyte term used to denote the typical sickle-cell deformity of patients with S hemoglobin.

Keratocyte red cells with a notch or several notches. Edges of the notches form projections that resenble horns.

Schizocyte red cells that represent fragments of red cells. The are small with angular sharp projections.

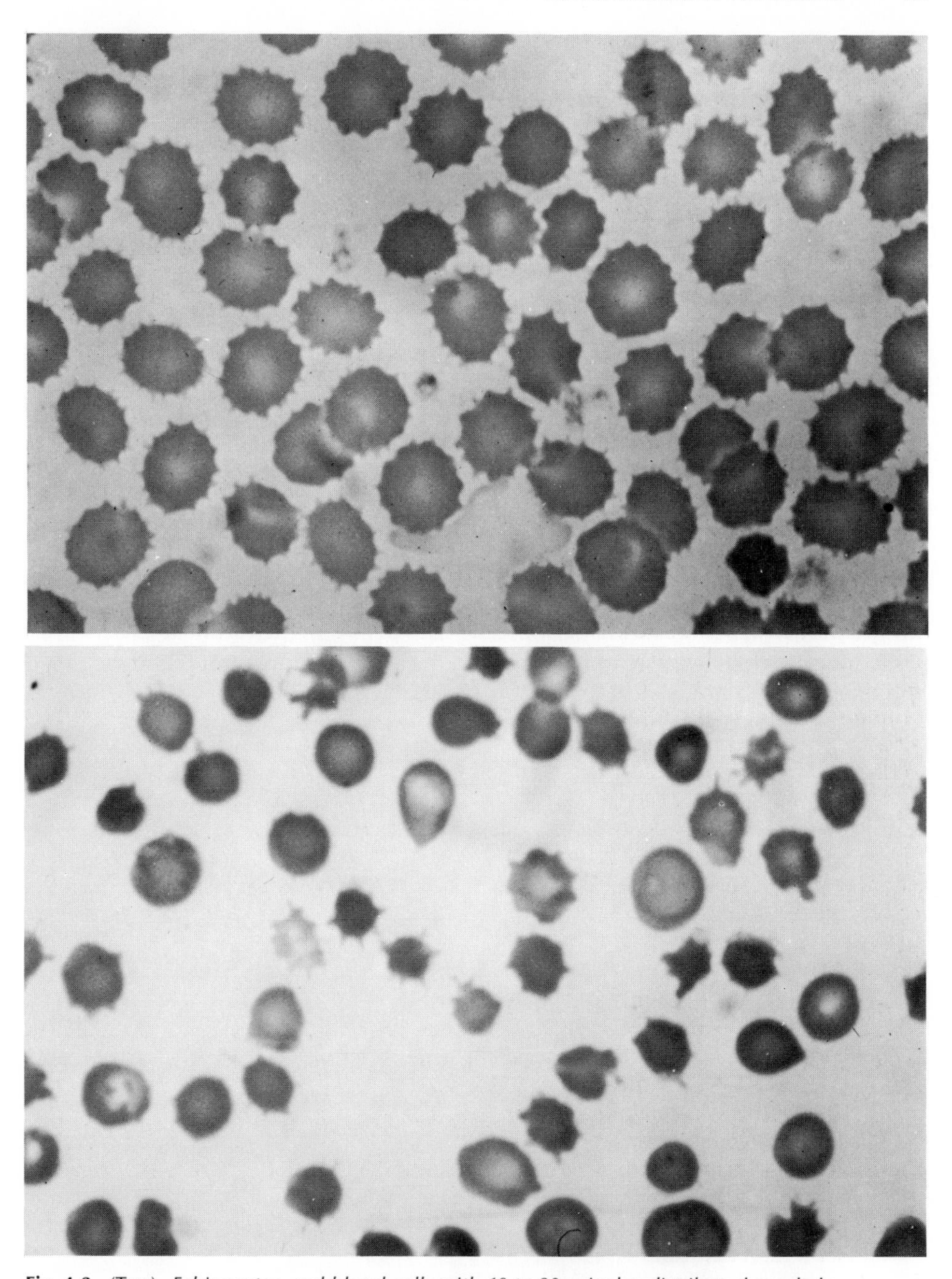

Fig. 4-2. (Top). *Echinocytes, red blood cells with 10 to 30 spicules distributed regularly over the surface. This striking picture is almost always artifactual in origin.* (Bottom.) *Acanthocytes, red blood cells that have a spheroidal shape with 2 to 20 spicules distributed irregularly over the surface.*

The spiculated red cells produced by exposure to EDTA and glass tubes belong to the category of echinocytes and acanthocytes. Striking echinocytes with a very uniform appearance (Fig. 4-2a) are invariably the result of artifactual change. The spiculated red cells of EDTA artifact usually are more a combination of echinocytes and acanthocytes (Figs. 2a, 2b). The spiculated red cells may be the result of glass exposure (alkaline medium) as well as EDTA effect because exposure of red blood cells to alkaline medium will result in echinocytic transformation of the red cells. The other forms of spiculated red cells are not seen secondary to EDTA. Stomatocytes are red cells with a mouthlike appearance to the central area of pallor, as shown in Figure 4-3. They may result from the exposure of red blood cells to an acid medium and to various chemical compounds. They are a common artifact seen in both EDTA smears and fingerstick smears.

We have tested the effect of EDTA concentration on red cell morphology in our laboratory. The EDTA evacuated tubes designed to collect 4 ml of blood were filled with 2, 4, and 5 ml of blood and stored for 1 hour at room temperature, and then blood smears were prepared. Blood samples were collected from ten normal subjects. The slides prepared from the evacuated tubes that contained 4 ml and 5 ml showed only minimal rouleaux, spiculated red cells, spheroidal red cells, and target cells, whereas the smears made from the 2 ml evacuated tubes showed marked rouleaux, target-cell formation, and increased numbers of spiculated red cells. Blood smears prepared from evacuated tubes with "short" samples appear to show much greater artifact than those prepared from tubes that contain the proper fill or slight excess fill.

Fig. 4-3. *Stomatocytes, red blood cells that have a central mouthlike appearance. Stomatocytes are a frequent artifactual finding.*

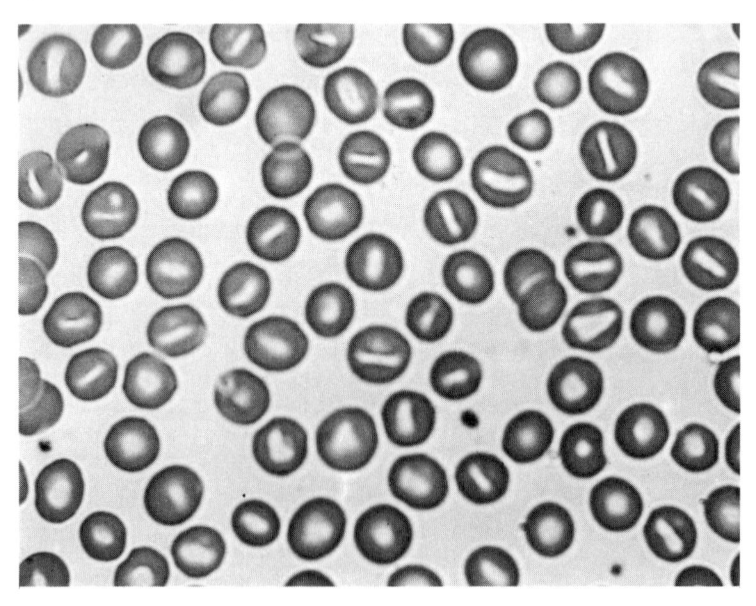

EFFECT ON LEUKOCYTES

The EDTA effect on leukocytes on smears prepared within 1 hour of collection is minimal. Occasionally vacuoles may appear in monocytes, granulocytes, and lymphocytes. In rare instances, EDTA-induced platelet satellitosis may be observed. Bloods stored in EDTA and kept at room temperature will show no measurable change in the white blood cell count in 24 to 30 hours as performed by the Coulter Model S Plus. The leukocyte differential done on the Technicon Hemalog D-90 system will also show no demonstrable change, but blood smears prepared from vials stored for 24 hours or longer will show marked deterioration of the cells with autolysis, vacuolization, and degranulation often to the point that recognition of the leukocytes on a Romanowsky-stained blood smear is nearly impossible. Polymorphonuclear leukocytes may develop hypersegmentation of their nuclei in stored blood, and some cells may show frank necrobiosis with karyorrhexsis of the nuclei.

In clinical situations in which the hematologist wishes to examine the red blood cell morphology critically, he will usually request a fingerstick blood smear to eliminate the potential source of the anticoagulant or the glass-induced artifacts described above.

PREPARATION OF THE BLOOD SMEAR

Various methods of preparation of peripheral blood smears are used in clinical laboratories at present. The oldest method, and still probably the most widely used, is the "wedge" smear method.

PREPARATION OF "WEDGE" BLOOD SMEARS

The blood films should be made on clean 25 mm × 75 mm (1 × 3 inch), 0.8 mm to 1.2 mm thick, glass microscope slides of good quality that are precleaned or cleaned in the laboratory before use. The films must be made within 1 hour of collection of the blood in EDTA. Fingerstick preparation should be made directly from the drop of blood on the finger that appears after the first drop has been cleaned away. The EDTA blood samples should be stored at room temperature (18–25°C) and should *not* be refrigerated. The sample should be mixed adequately before preparation of the blood film. A drop of blood sufficiently large to produce a blood smear of at least 2.5 cm should be placed near one end of the microscope slide. A second "pusher" slide or similar object is held at a 45° angle to the slide. The "pusher" is drawn backward until it touches the drop of blood. The blood will spread behind the pusher slide by capillary attraction and should be allowed to spread to full width of the pusher slide. The pusher is then advanced rapidly to produce a blood film at least 2.5 cm long and ending at least 1 cm before the end of the slide (Fig. 4-4). The end of the blood film should be straight and not bullet shaped. (*see* Fig. 4-5). A bullet-shaped blood film results from advancing the pusher slide before the drop of blood has spread uniformly behind the pusher slide. The straight-edge blood film is preferred to the bullet-edge film because the leukocytes are distributed more uniformly. A narrow pusher slide that produces a blood film that does

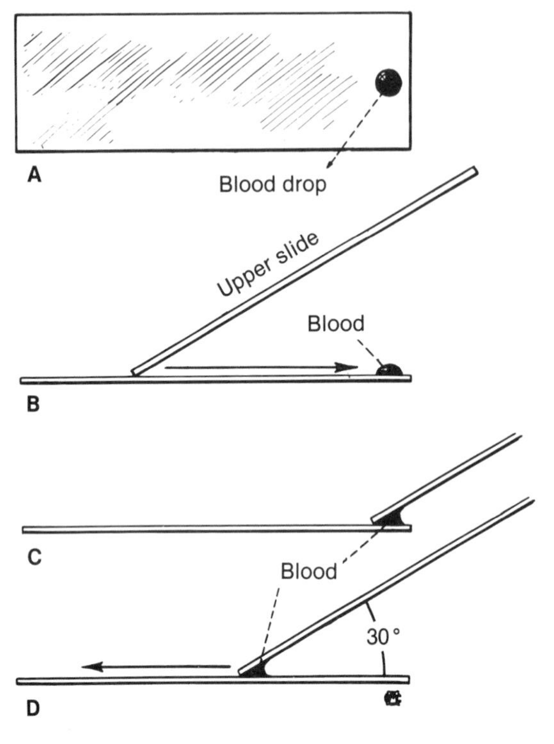

Fig. 4-4. *Technique for preparation of "wedge" blood smears.*

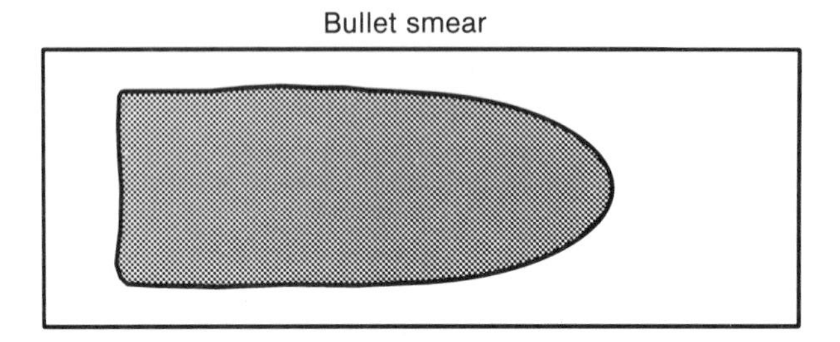

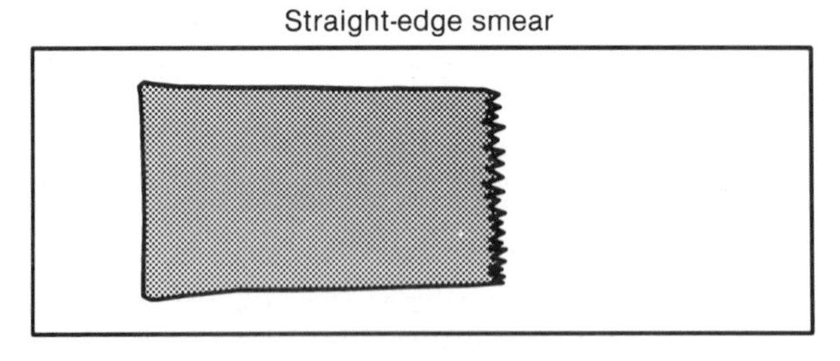

Fig. 4-5. *Bullet and straight-edge smears are the product of methods of blood smear preparation.*

not extend to the edge of the glass slide is also preferred; if a pusher is used that is the same width as the slide, there is a greater accumulation of leukocytes at the edges of the film. Blood films of good qualtiy are extremely difficult, or in some cases impossible, to make from patients with extremely high or low hematocrits. Good-quality smears with good distribution of red cells in the examination area of the slide may also be difficult or impossible to prepare in patients with cold agglutinin disease, cryoglobulinemia, monoclonal gammopathies, or in the presence of some drugs such as miconazole.

Blood films must be dried quickly to obtain good red blood cell morphology. If the blood film dries slowly in a humid environment, a "punched out" artifact of the red cells may result, leading to an erroneous interpretation of hypochromia of the red cells. Blood films should be fixed within 1 hour of preparation if a delay in staining of the smears is anticipated. The slides may be dipped in absolute (less than 3% water) methanol, which preserves the staining quality of the smears. Heat should never be used to dry blood films rapidly because it will cause shrinking of the blood cells and make their identification difficult.

MECHANICAL, SEMIAUTOMATED, AND AUTOMATED PREPARATION OF "WEDGE" SMEARS

The preparation of good-quality "wedge" blood films requires some dexterity, skill, and practice. Routine production of good-quality smears by venipuncture technicians who do not perform them regularly is the exception rather than the rule. The introduction of automated instruments to perform leukocyte differential counts on "wedge" smears has made the preparation of good-quality blood films even more important. The automated instruments require better quality smears to perform well. The human observer can often adapt somewhat to irregularities in smear preparation that would make use of the smear on an automated instrument impossible. Various mechanical devices have been proposed and marketed over the years to prepare "wedge" blood films, but the only commercial device in wide use at present is the "Hemaprep" device (Geometric Data Corporation). This compact instrument can make acceptable "wedge" smears for use on the Hematrak instrument from most patients. Patients with abnormalities that prevent preparation of manual "wedge" preparations will also affect the performance of the Hemaprep. The instrument is adjustable to compensate for high or low hematocrit examples

A fully automated method of preparation of "wedge" smears has been introduced: the Autoslide, a device installed on the Technicon Hemalog D instruments or the Technicon H6000 instrument (Fig. 4-6). The device makes a wedge smear from blood dropped onto a nylon mesh. The blood is smeared onto a plastic tape and then dried, and the tape is passed through a staining bath as shown in Figure 4-6. Identification number, date, and sequence number are printed onto the tape. An acrylic monomer is placed on the tape and a glass slide added ("coverslipping" the tape). The blood film is then placed between the plastic tape and the glass slide. The acrylic monomer is cured by an ultraviolet light source. The tape is peeled away, leaving the blood smear with its identification transferred to the glass slide. The "autoslides" have the appearance of wedge smears with a thick end, a suitable working area, and a thin end. They may be examined by a conventional microscope or examined on the Hematrak system with suitable adaptation of the Hematrak stage mechanism.

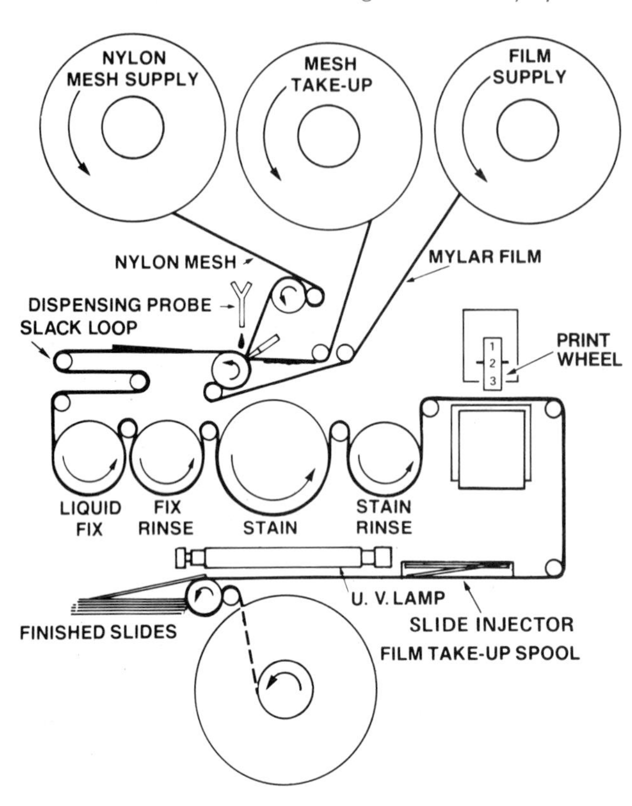

Fig. 4-6. *Diagrammatic representation of the Technicon Autoslide. The Autoslide is attached to the Technicon Hemalog D or H6000 systems. A blood film is prepared from an aliquot of each sample aspirated by the instrument for performance of a differential leukocyte count or in combination with a complete blood count. The blood film is prepared, stained, dried, labeled, "coverslipped," and ready for examination as it comes off the instrument.*

Another method to prepare blood films is the use of centrifugelike devices, such as the Larc spinner (Corning), the Slide spinner system used with the Coulter Diff-3, and the Abbott ADC-500, which spin the slide in a centrifuge to produce a monolayer of red cells with uniform distribution of leukocytes and platelets. The Larc spinner uses EDTA anticoagulated blood and relies on an optical sensing device to determine the density of the film to control the spinning time and obtain uniform-density blood films. The Abbott blood film centrifuge also uses EDTA anticoagulated blood. Because the viscosity of blood largely depends on the hematocrit of the blood and the viscosity increases markedly with hematocrits above 55 but is relatively constant between 0 and 50, as seen in Figure 4-7, the Coulter Diff-3 system uses diluted blood to prepare spun blood films. The dilution is performed automatically when the sample is aspirated into the probe of the centrifuge device. The advantages of the centrifuge-prepared blood films are their uniform density of red cells and uniform distribution of leukocytes and platelets on the film. These spun blood forms provide a much larger working area in which the instrument can perform the differential count and eliminate some errors owing to maldistribution of leukocytes on the film. The disadvantages of the blood-film centrifuge devices is that they cannot use direct fingerstick samples at the bedside and must use anticoagulated blood samples transported to the laboratory.

Another method of preparation of peripheral blood films that can produce excellent-quality preparations but is time consuming and suitable only for low-

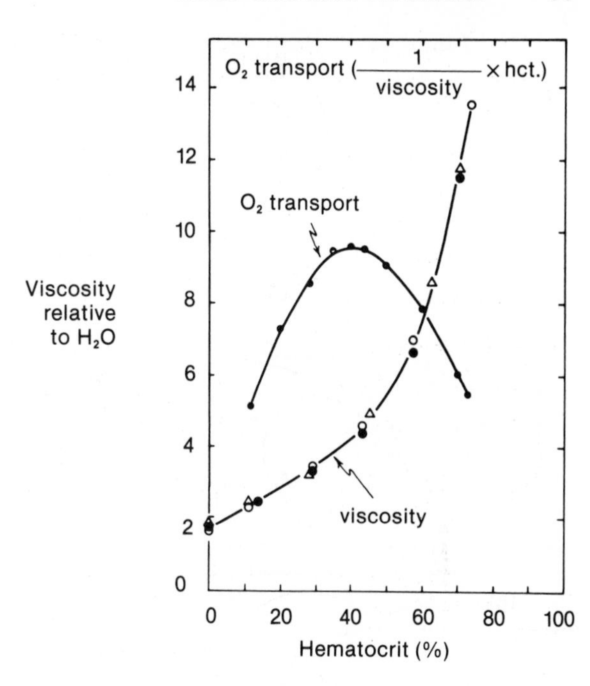

Fig. 4-7. *The effect of hematocrit on blood viscosity and oxygen transport. (Erslev A, Gabuzda T: Pathophysiology of Blood, p. 31. Philadelphia, WB Saunders, 1975)*

volume specialized applications is the "coverslip" method. The coverslip method uses two coverslips: A drop of blood is placed on one coverslip, and the second coverslip is placed on top. The blood is allowed to spread between the coverslips into a thin film. The coverslips are then pulled apart in a horizontal sliding motion, and the film is dried quickly. The small delicate coverslips are difficult to handle and must be stained by manual methods.

REFERENCE

1. Bessis M: Blood Smears Reinterpreted, p 25. (Translated by Brecher G.) Berlin, Springer International, 1977

Blood Bank Specimens

Cheryl L. Sonnenberg

The principal function of a phlebotomist is to collect blood specimens that will be tested in various clinical laboratories. The procedures performed in those laboratories yield information that the physician uses to diagnose, monitor, and treat illnesses.

The blood bank laboratory is unique in that it provides products for the management of patient therapy. Blood bank personnel perform procedures that yield information indicating whether blood products can be transfused safely to a patient. If the scientific basis of blood banking is understood and strict procedures are followed to identify a patient, a specimen, or a unit of blood correctly, a possible hemolytic transfusion reaction may be forestalled.

IMMUNOLOGY AND IMMUNOHEMATOLOGY FOR THE PHLEBOTOMIST

Immunology is the study of the immune system, the body's means of defense. The task of the immune system is to destroy and to eliminate substances and invasive organisms such as bacteria that may harm the body.

The composition of these substances and invasive organisms includes molecular structures called antigens. These antigens, which are foreign or unknown, stimulate the immune system to respond in one of two ways: In *cellular response,* white blood cells called T lymphocytes are stimulated to play an active role in the destruction and elimination of foreign substances and organisms. This mechanism is involved in rejection of transplanted tissues, such as kidneys; and in *humoral response,* a protein called antibody is synthesized by plasma cells, one type of white blood cells. Once antibody is produced, it is released into plasma and other body fluids. If the antibodies come in contact with the specific antigen that caused its production, a reaction will take place forming an antigen/antibody complex (Fig. 4-8). This complex is carried by the blood stream to the spleen, which has an active role in recognizing antigen/antibody complexes on foreign substances and organisms, as well as destroying and eliminating them.

Immunohematology is the study of red blood cell antigens and antibodies. More than 400 different red blood cell antigens and their specific antibodies have been identified. The antigens present on a person's red blood cells depend on inherited characteristics from each parent. Some antigens are present on almost everyone's red blood cells, some are present only rarely, and others are present in various percentages.

Antibodies are occasionally produced in response to exposure of foreign red blood cell antigens through blood transfusion or pregnancy. If a patient with an antibody is transfused with a blood product containing the specific antigen, a transfusion reaction will occur, resulting in destruction of the transfused red blood cells. Destruction of red blood cells can occur by means of intravascular or extravascular hemolysis. *Intravascular hemolysis* is an immediate destruction of transfused red blood cells within the blood stream. The circulating antibody comes in contact with the specific red blood cell antigen, forming an antigen/antibody complex. This complex initiates the activation of another protein called *complement.* The result is the leakage of hemoglobin from within the red blood cell, and thus destruction of the red blood cell. Destruction of the red blood cells by this mechanism is called *hemolysis* (Fig. 4-9). The complexes also activate other proteins that function in the blood clotting process. Blood clots are trapped in the minute tubules of the kidney and other small capillaries throughout the body. If this process cannot be stopped, the patient may die.

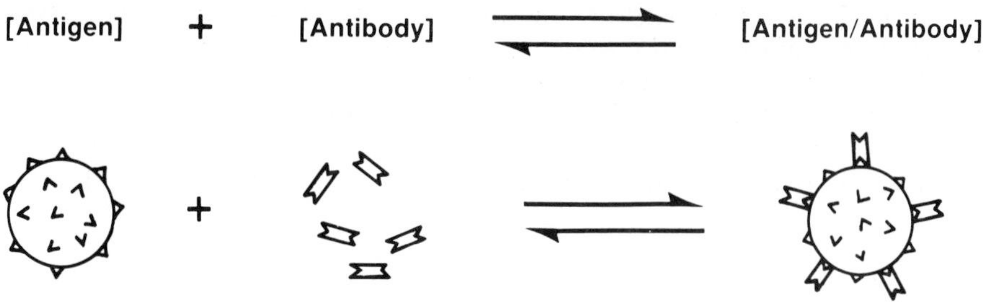

[Antigen] + [Antibody] ⇌ [Antigen/Antibody]

Fig. 4-8. *Formation of antigen/antibody complex.*

Fig. 4-9. *Hemolysis.*

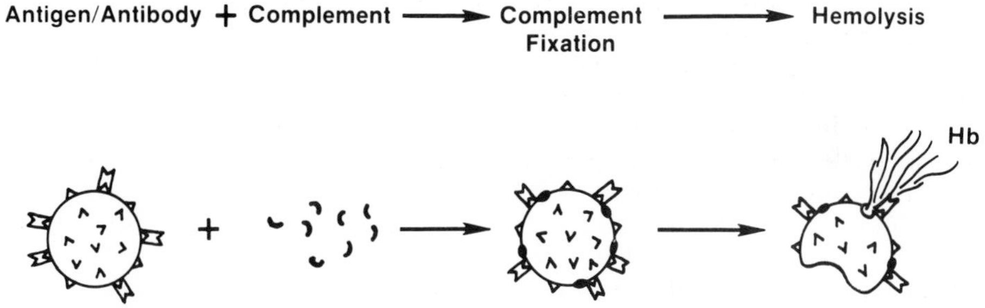

Antigen/Antibody + Complement → Complement Fixation → Hemolysis

Hb

Extravascular hemolysis is usually a delayed process that shortens the lifespan of the transfused red blood cells over time. The antibody is frequently at very low levels within the plasma and may not be recognized by procedures used in the blood bank laboratory. When red blood cells containing the specific antigen are transfused, the plasma cells are restimulated and begin producing antibody. As the circulating antibody comes in contact with the antigens on the red blood cells, an antigen/antibody complex is formed. The spleen filters out these red blood cells with antibody attached. White blood cells called *macrophages* engulf the red blood cells and destroy them (Fig. 4-10). The patient may have a transfusion reaction that varies in severity depending on the particular type of antibody, as well as on the amount of antigen and antibody present.

Red blood cell antigens and antibodies are grouped according to their characteristics into blood group systems. The most important of these blood group systems is the ABO system, unique in that ABO antibodies are present in the liquid portion of blood and plasma throughout most of the patient's life span. The antibodies are thought to be formed in response to exposure to bacteria, plants, and other environmental factors that have antigens similar to the ABO antigens of red cells. A person who has type A blood and A antigen on his red blood cells, for example, will be exposed to B antigen early in life. The B antigen is foreign to that person, and B antibody will be produced in response to that stimulation. Persons with type O blood have neither A nor B antigen on the red blood cells; therefore A and B antibodies will be produced. The same situations will apply to the other blood types (Table 4-1).

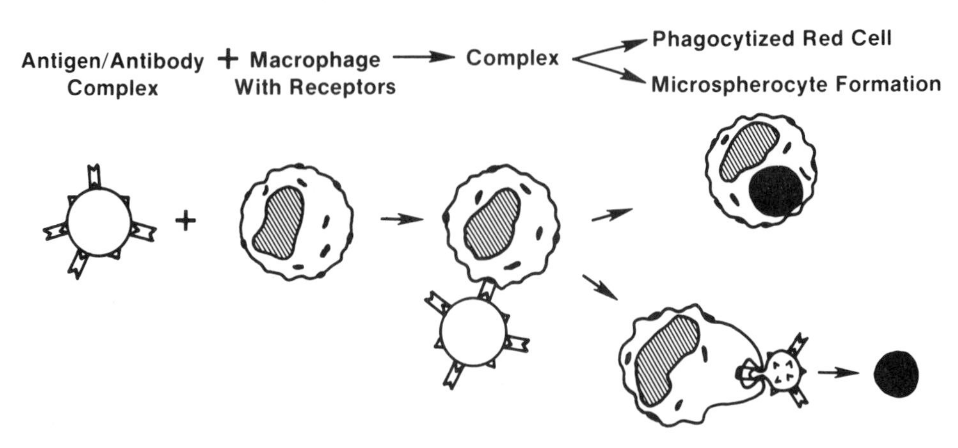

Fig. 4-10. *Macrophages engulf red blood cells, destroying them.*

Table 4-1. *ABO System*

BLOOD TYPE	ANTIGENS	ANTIBODIES	PERCENTAGE IN WHITE POPULATION
A	A	Anti-B	40
B	B	Anti-A	11
AB	A and B	None	4
O	None	Anti-A and Anti-B	45

The ABO antibodies are hemolytic, causing hemolysis of red blood cells. The antibodies are abundant in a person's plasma. If a patient with B antibody is transfused with type B red blood cells, an immediate intravascular hemolytic transfusion reaction will occur as a result of incorrect identification of a patient, a blood specimen, or a unit of blood. A physician orders, for example, a blood transfusion for patient X. The phlebotomist accidentally collects the blood specimen from patient Y, labeling the specimen with patient X's identification information. The blood specimen is tested in the blood bank laboratory and found to be type A. Type A blood is issued for transfusion. After receiving the unit of blood, patient X has an immediate hemolytic transfusion reaction. On investigation, it is found that patient X has, in fact, type B blood. His A antibody hemolyzed the type A red blood cells that were transfused.

ERRORS IN BLOOD TRANSFUSION THERAPY—FROM VENIPUNCTURE TO TRANSFUSION

Transfusion of blood is prescribed for patients as a treatment for certain medical conditions. As with any drug, safe administration of blood is essential. Strictly adhering to protocols will aid in avoiding harm to the patient during transfusion therapy. A number of situations that may harm the patient are listed below.

1. Specimen collected from wrong patient
2. Specimen labeled with incorrect identification

3. Incorrect laboratory test result or report, that is, incorrect ABO/Rh type
4. Blood or blood product mislabeled
5. Patient transfused with wrong blood or blood product

Situations 1 and 2 involve the phlebotomist directly. It may seem inconceivable to a phlebotomist that a blood specimen could be collected or labeled incorrectly, and yet these situations do occur.

In September 1980, Myhre reported on 113 cases of transfusion-related fatalities occurring from April 1976 through 31 December 1979.[1] These cases were originally reported to the Food and Drug Administration, Bureau of Biologics (FDA–BoB).

Of the 113 cases reported, 47 cases were due to clerical errors resulting in ABO incompatibilities between the recipient and the transfused donor unit. A *clerical error* can be defined as an incorrect comparison or transcription of information from one record to another, that is, wristband and specimen label.

In seven of these cases, the blood specimen collected for ABO/Rh typing and cross-matching was mislabeled or collected from the wrong patient. Nine cases involved clerical error in the laboratory; in 30 cases the wrong unit of blood was transfused to the patient; and in 1 case, a mix-up occurred between two units of blood placed in the blood warmer. Myhre stated that these transfusion fatalities could have been prevented if established protocols had been followed.

No one makes mistakes intentionally, and yet errors do occur. How does an error occur during the collection and labeling of blood specimens? Three questions need to be asked when analyzing this problem.

- Are protocols written in a clear and easy-to-follow manner?
- Are phlebotomists following each step within the established protocol?
- Is the phlebotomist encountering unusual situations not described in the protocol?

If a phlebotomist makes an error, the cause may be related to one or more of these items.

WRITING PROTOCOLS

Accurate and useful protocols are essential to ensure proper collection and labeling of blood specimens. Protocols should be written so as to describe clearly and simply the step-by-step procedure. The steps within the procedure should be in order of the sequence of events.[3]

An approach that lends itself well to phlebotomy procedures is the responsibility/action format (Table 4-2), particularly useful when a protocol includes more than one group of people. It outlines clearly who is responsible for each activity.

Once the protocol has been written and approved by the staff, copies should be distributed and reviewed with the persons involved. The protocols must then be evaluated to ensure the established system is workable, and the purpose is fulfilled.

Table 4-2. *Example of Responsibility/Action Procedure Format*

RESPONSIBILITY	ACTION
Physician	1. Complete the blood transfusion request card with the following information: a. patient's identification number; b. patient's full name; c. blood product requested; d. number of products requested; e. time/date of transfusion; f. signature. 2. Send request to venipuncture laboratory.
Phlebotomist	3. Enter venipuncture request on day sheet.

FOLLOWING ESTABLISHED PROTOCOLS

Well-written protocols are especially valuable in training new personnel. A protocol delineating every step of the procedure aids the new employee in performing the procedures correctly. This type of protocol also ensures that no shortcuts are taught in the training period.

Even with written protocols and training programs, errors can and do occur. Some of the ways to avoid errors are as follows.

Phlebotomist should be trained properly.
Training program should be thorough.
Phlebotomist should be informed of changes in protocols.
Protocols should be written clearly.
Established protocols should address resolution of problem situations.

Mechanisms should be established that minimize and monitor the occurrence of errors. One good way of achieving this is by establishing a periodic continuing education program to review the significance of following established procedures, to describe changes in protocols, and to establish a forum for discussion of problems that need to be resolved. The phlebotomist is the first individual to discover a *weak link* in protocols. Once a problem is identified, resolution must be considered and instituted.

Another means of minimizing errors is the establishment of standards. Standards have been set by the American Association of Blood Banks (AABB) to help ensure proper patient and specimen identification. If a blood bank facility is a member of the AABB, it is required to comply with these standards. Compliance is monitored by periodic inspections.

Two AABB standards pertinent to this discussion are listed below.

E.2.100 Standard for Labeling of Blood Specimens[2]

The intended recipient and the blood sample shall be positively identified at the time of collection. Blood samples shall be obtained in stoppered tubes identified with a firmly attached label bearing at least the recipient's first and last names, identification number, and the date. The completed label shall be attached to the tube before leaving the side of the recipient, and there must be a mechanism to identify the person who drew the blood.

E.2.200 Standard for Identifying Information on Blood Sample[2]

Before a specimen is used for blood typing or compatibility testing, a qualified person in the blood bank or transfusion service shall confirm that all identifying information on the request form is in agreement with that on the specimen label. In case of discrepancy or doubt, another specimen shall be obtained.

It is essential for phlebotomists and blood bankers to work together and follow these standards to ensure that all patients receive the safest blood transfusion therapy possible.

UNUSUAL SITUATIONS ENCOUNTERED BY THE PHLEBOTOMIST

Protocols are often written to describe an ideal system. The author of the protocol assumes that the system will always work. It is important to realize that there are often breakdowns in the system because of unusual or unexpected circumstances. Thus, whenever a protocol is written, one must always address these situations by delineating steps to be taken in unusual circumstances. Some of these unusual situations are discussed below.

NO WRISTBAND FOR PATIENT ON HOSPITAL FLOOR

Although all patients should have wristbands in place, sometimes the wristbands may be missing. The patient may be newly admitted or may have had the wristband cut off because it interfered with a procedure.

The phlebotomist may find that the wristband has been removed and taped to the bed or table. These wristbands are *not* acceptable for use. The wristband may not belong to the patient in the bed but to a patient who previously occupied that bed.

A procedure should be established to ensure that a wristband is generated and placed on the patient. In an emergency, the phlebotomist must have an identification protocol to follow for proper identification of the patient. If the patient is coherent, for example, the phletobomist should have the patient *state* and *spell* his full name. The identification numbers should be transcribed from an available record. If the patient is incoherent or unconscious, a third party (floor nurse or family member) should identify the patient by stating and spelling the patient's full name. *A word of warning:* Misidentification of a patient has occurred in this situation, especially when there has been more than one patient in the room.

NO WRISTBAND FOR PATIENT IN THE EMERGENCY ROOM

An area where a tragic mishap may occur is in the emergency room, particularly if it is a trauma center. Frequently, patients cannot be identified at admission, and thus the usual identification protocol cannot be used.

Excellent commercially prepared identification systems are available for use in the emergency room. This system of identification uses letters and numbers on a temporary wristband as well as labels for specimens and units of blood to be transfused; for example, the patient wristband, blood specimen, and units of blood to be transfused are identified as FBC 654. If the attending physician collects the blood specimen in a syringe from a patient, it should be labeled with an alpha-numeric label to ensure no specimen mix up.

An example of a particularly confusing situation in the emergency room is the admission of several family members with the same last name. It is not unusual to have a father who is *senior* and a son who is *junior.* These patients are at greater risk in blood transfusion accidents if not identified very carefully.

INCORRECT INFORMATION ON PATIENT WRISTBAND

Wristbands can be generated with incorrect information on them. Unless all information is correct on all records used for identification of the patient, transfusion therapy cannot begin.

The phlebotomist may not be aware of an incorrect wristband because the name or number on the phlebotomy request may also be incorrect. Blood bank personnel frequently recognize these errors because historical ABO/Rh typing records are always compared with current information. A protocol should be written to ensure that proper corrective action takes place when incorrect wristbands are discovered.

PATIENTS WITH SIMILAR OR SAME NAMES IN HOSPITAL

It is not unusual for two patients with a similar or the same name to be in the hospital. The major difference in the identification of these patients is the assigned hospital number. If close attention is not given to identification protocols, the blood specimen may inadvertently be collected from the wrong patient or the patient may be transfused with the wrong blood.

With an increase in tissue transplantation procedures, the phlebotomist will see more family members in the hospital; the recipient and the donor may have a similar or the same name. Multiple births, particularly twins, are not uncommon, and frequently these newborns are admitted to the pediatric intensive-care unit. The immediate identification of these newborns may be Baby Boy Smith #1 and Baby Boy Smith #2. The phlebotomist should take great care in the identification of these patients.

A mechanism should be established to alert phlebotomists when patients with similar or the same names are admitted to the hospital. Strict adherence of established identification procedures is essential for patient safety. The phlebotomist has a vital role in the collection of patient specimens used by blood bank personnel to provide blood transfusion therapy. Fortunately, transfusion mishaps are infrequent, but because the dramatic results of these mishaps can be the death of the patient, special emphasis must be placed on this task. Phlebotomists should have a sound understanding of the scientific basis of blood banking and must have thorough training in performance of their assigned duties. Protocols must be written clearly and precisely to aid the phlebotomist in carrying out the assigned function. Adherence to those procedures is essential. The phlebotomist should be informed immediately of any procedural changes and must inform the staff of any problems that occur to correct potentially dangerous situations.

BIBLIOGRAPHY

Myhre B: Fatalities from blood transfusion. JAMA 244:1333–1335, 1980

Committee on Standards: Standards for Blood Banks and Transfusion Services, 9th ed. Washington DC, American Association of Blood Banks, 1978

Matthies L: The Playscript Procedure: A New Tool of Administration. New York, Office Publications, 1961

Part II

COLLECTION OF NONBLOOD SPECIMENS

5

Specimens for Urinalysis

Jean Simindinger
Fuad K. Mansour
Jean M. Slockbower

Urine is formed in the tubules of the kidney by filtration and selective reabsorption and then stored in the bladder. Normally more than 1000 liters of blood are filtered by the kidneys each day, resulting in over 1 liter of urine per day. The body uses this system to regulate the concentration of certain substances in the blood, to eliminate wastes, and to regulate the amount of water in the body.

Urine specimens are used in the laboratory to diagnose and to manage renal or urinary tract diseases and to detect metabolic or systemic diseases. The various methods and timings of urine specimen collections depend on what tests have been requested by the physician.

SINGLE SPECIMENS

A random urine specimen collected in an unsterile container usually is used for routine urinalysis to analyze the chemical content of a specimen for such variables as glucose, protein, *p*H, specific gravity, osmolality, color and possible presence of blood, pus, or crystals. Often the first morning-voided specimen is requested because it gives the urine concentration most accurately, but otherwise the time of collection or volume of the voiding is not essential. Prompt transport of the specimen to the laboratory remains essential. If a delay is inevitable, placing the specimen on ice may help preserve the freshness of the sample. These types of specimens are never used for the purpose of culturing. Containers for routine random specimens should be chemically clean, should hold about 50 ml in volume, and must have a tight-fitting lid to prevent leakage during transportation. The container itself, not the lid, must be labeled with identification information, and the label must be able to adhere to the container if the specimen is refrigerated.

STERILE CLEAN CATCH

The object of a "clean-catch" voided urine collection is to obtain approximately 60 ml of urine from a person's bladder and urethra in a sterile container. Bladder catheterization has not generally been replaced by the clean-voided midstream urine technique that eliminates the risk of introducing infection. The best laboratory techniques for counting and identifying bacteria are of little value if the specimen is not collected properly and transported appropriately. The following offers a step-by-step guide as to how to carry out this very frequent doctor's order: collect urine specimen for culture.

1. The patient must realize the importance of cooperation. If the patient understands that a carelessly collected urine specimen could give inconclusive results that necessitate a repeat specimen, thus delaying treatment, or could cause the wrong treatment to be given or additional costs to be borne, then the patient is more likely to follow good collection technique directions. Stimulating the patient's interest in his own health, and decreasing his anxiety with both verbal and written instructions, will aid in successful specimen collection.
2. The cleansing procedure must remove contaminating organisms from the vulva urethral meatus, and related perinal area, so that bacteria found in the urine specimen can be assumed to come from the bladder and urethra only. Extraneous bacterial contamination of urine comes from the following sources.

    - Bacteria from the hands, skin, or clothing may enter the collecting vessel.
    - Hair from the perineum may fall into the collecting receptacle.
    - Bacteria from beneath the prepuce in males may contaminate the stream.
    - Bacteria from vaginal secretions or distal urethra in females may contaminate the steam.

A specimen taken from a bedpan, a break in a drainage system, or from a drainage bag for a 24-hour urine collection cannot be cultured.

FEMALE PATIENTS

The patient should have a strong urge to void. Her privacy should be assured. She should then perform the following steps.

1. Wash hands before and after voiding.
2. Remove all clothing below the waist and have all equipment ready.
3. Remove cap from sterile wide mouthed jar and put aside, with inside of jar lid up. Never touch the rim of the jar while handling the jar.
4. Using five sterile 4-inch × 4-inch gauze pads soaked with 10% solution of liquid antiseptic soap, begin wiping the labia from front to back using each of the five sterile pads once, while spreading the labia with the other hand. Begin wiping outer labia first and work toward middle, using the last pad down the midline or over urinary meatus. Remember, the *friction* of cleansing more than the soap prepares the collection area.

5. Using five sterile *dry* 4-inch × 4-inch gauze pads, repeat the procedure again while remembering to keep the labia spread.
6. Void a small amount, stop, and then begin voiding directly into collection cup, until cup is half full.
7. Carefully seal the cap of the container so that it is tight and leak proof.
8. Finish voiding.

MALE PATIENTS

The patient should have a strong urge to void. His privacy should be assured. He should then perform the following steps.

1. Wash hands before and after voiding.
2. Remove or open any clothing to facilitate voiding and have all equipment ready.
3. Wash penis with several gauze-soaked antiseptic soap pads, using one each time, ending the last wipe near the tip of the penis. Uncircumcised males should retract foreskin to wash glans.
4. Repeat with several dry gauze pads, used separately.
5. Collect the specimen the same as for instructions for females.

The specimen should have the patient's name and other significant information such as time placed on the specimen cup, not the lid, and must be delivered promptly to the laboratory in an upright position with a requisition form. Specimens that cannot be transported immediately should be on ice and delivered as soon as possible. A first-morning urine sample, after overnight "bladder incubation," is ideal for testing.

A bedfast patient will need the assistance of one or two adequately trained persons to cleanse him and collect the specimen. For females, a strong urine flow is needed to obtain a midstream specimen from a patient on a bedpan, as the urine collected must never have first flowed over the anal area.

Infections usually are associated with bacterial counts of 100,000 (10^5) or more organisms per milliliter of urine. Urine is an excellent culture medium for most organisms that infect the urinary tract, and growth occurs in the urine itself *in vivo,* resulting in high counts because of untreated infections. If contamination of the urine is from the external genitalia, in the absence of infection, the count of organisms is usually 1000 (10^3) per milliliter of urine in properly collected and transported specimens.

In women, urinary tract infections commonly do not begin in the kidneys but in the urethra and urinary bladder. The kidneys may become involved, and 90% of cases of child-bearing age are caused by *Escherichia coli.* Other organisms including *Klebsiella* species, *Enterobacter* species, *Proteus* species, *Pseudomonas* species, and enterococci are frequently encountered in patients with obstructive lesions, in those with paralytic disease affecting the bladder function, or in those who have had manipulation of the urinary tract. Immunodeficient patients or patients with uncontrolled diabetes commonly may have *Candida* species involved in urinary tract infections. *Staphyloccocus* and *S. epidermidis* can also cause cryptitis.

Eighty percent of male patients diagnosed as having bacterial prostatitis have gram-negative bacilli, with *E. coli* as the cause. Often two or more kinds of bacteria may be present, including *Klebsiella* species, *Proteus mirabilis, Pseudomonas* species, and *Enterobacter* species. *Enteroccus,* a gram-positive bacteria, occasionally causes bacterial prostatitis.

TIMED SPECIMENS

Types of timed specimens include tolerance specimens where urine is collected in a series of bottles at designated times; 2- and 3-hour specimens, with some done during certain hours and others during certain times after meals; and the more traditional 12- and 24-hour collections. The following factors should be remembered for any of these urine collections.

- Collect the specimens properly and carefully.
- Instruct the patient on how to collect each voiding and on the time element (when the test begins and ends). The first specimen is always discarded, and all other voidings are collected, including the last voided specimen.
- Obtain enough correct, nonleaking vessels for collection and add the correct preservatives, if needed. Note on container what preservative was added.
- Refrigerate the entire volume during collection.
- Label the containers with the patient's name, patient information, and times and dates of collection. An intake volume may also be needed.
- Transport to the proper laboratory promptly with requisition form.
- Note any aliquot amounts that have been removed or volume totals of discarded specimens on requisition.

INSTRUCTION PROCEDURES

The following procedures may be used for instructing a patient on his timed urine collection.

1. Seat the patient in a private area or instruction room.
2. Inform the patient of the urine collection. Explain and review in detail the procedure he should follow and give him a printed copy of the instructions so that he may read them at the same time and refer to them later during collection (*See* example on p. 108, Patient Instructions for Fractional Glucose Test.)
3. Ask appropriate questions that apply to the test information.
4. Fill out all test cards and urine bottle labels. Tape label to collection bottle.
5. Give patient the container placed in an outer wrap, such as a plastic bag with draw string, that is easily carried during the collection interval.
6. If it is appropriate to start the collection during the interview, instruct the patient to use the toilet and void. Then begin to time the collection.

An example of this type of collection with instructions that can be started during the interview is the 2-hour collection for amylase testing.

INSTRUCTIONS TO PATIENT FOR COLLECTING SPECIMEN FOR URINE AMYLASE TEST

________	—— Enter day of month collected. Specimens may be collected at any hour.
________ am ________ pm	___ Void urine completely in toilet. Record exact time.
exactly 2 hours	—— Immediately after voidng urine, drink 2 full glasses of water, and then drink nothing more for the next 2 hours.
________ am ________ pm	—— During and at the end of this 2-hour period, void all urine in specimen container. Record exact ending time.
	Return capped container to desk.

Some urine collection procedures may include instructions for a period of time preceding the collection. The following is an example of the instructions for 5-hydroxyindoleacetic acid testing.

INSTRUCTIONS FOR COLLECTING OF URINE SPECIMENS FOR 5-HYDROXYINDOLEACETIC ACID FOR THE 24-HOUR PERIOD PRECEDING URINE COLLECTION

- Do not eat avocados, plums, walnuts, pineapples, or eggplant.
- Do not use cough syrup containing glyceryl guaiacolate.
- Do not take Tylenol (acetaminophen) or Empirin (phenacetin) or any compound containing these drugs 24 hours before the test or during the collection.

24-HOUR COLLECTIONS

Because of the considerable time these collections take, it is important to review the details of this collection with the patient. An example of printed instructions useful to the patient is given below.

IMPORTANT
PLEASE READ BEFORE STARTING COLLECTION OF URINE

The accurately timed urine collection that you are about to make is an important part of your examination. Decisions important to your health may depend on it. This test is valid ONLY if the collection includes all urine that you pass in a 24-hour period. If for any reason some of the urine passed during this 24 hours is not put into the containers for the collection, the test will be inaccurate and may have to be repeated, requiring another day of your time.

Start the 24-hour collection period exactly at 7 a.m. Empty your bladder at that time and discard the urine because this urine was formed before the collection period. Thereafter collect all urine that you pass for the next 24-hours, until 7 a.m. the following morning. Exactly at 7 a.m., again empty your bladder to conclude the collection, but this time save the specimen because this urine was formed during the collection period. Should you have a bowel movement during the 24-hour period, try to pass your urine before the bowel movement so as to avoid loss of urine.

CONTAINERS FOR TIMED COLLECTION

Important factors in selecting containers are the size, a tight-fitting lid, and a wide mouth. A 1-liter bottle is adequate for 2-hour testing collection; a 2-liter bottle is adequate only for collection of glucose fractionations; and a 3-liter bottle is adequate for 24-hour collection. Two liter bottles should not be used for timed collections because one bottle will often not hold all the collection and a second bottle is needed. This second bottle may not arrive in the laboratory until well after the first bottle has been aliquoted and tested, thus resulting in inaccurate results.

PATIENT INSTRUCTIONS FOR FRACTIONAL GLUCOSE TEST

This test involves **four** urine collections for a 24-hour period. The schedule below should be followed for accurate test results.

Sixty minutes before breakfast empty bladder and discard urine. After discarding first urine, collect all urine passed in the 24-hour collection period.

I. Place all urine tested and collected in bottle #1.

URINE SPECIMEN

30 minutes before breakfast test this specimen to start the first collection. Indicate time started. Collect all urine that you pass into bottle #1.

EAT BREAKFAST

URINE SPECIMEN

60 minutes before midday meal collect specimen in first container. Indicate time ended. This ends the **FIRST** collection period.

II. Place all urine tested and collected in bottle #2.

URINE SPECIMEN

30 minutes before midday meal test this specimen. Include in container for second collection period. Indicate time started. Collect all urine that you pass into bottle #2.

EAT MIDDAY MEAL

EAT AFTERNOON SNACK (optional)

60 minutes before evening meal collect specimen. Indicate time ended. This ends the **SECOND** collection period.

III. Place all urine tested and collected in bottle #3.

URINE SPECIMEN

30 minutes before evening meal test this specimen. Include in container for third collection period. Indicate time started. Collect all urine that you pass into bottle #3.

EAT EVENING MEAL

URINE SPECIMEN

60 minutes before bedtime snack collect specimen. Indicate time ended. This ends the **THIRD** collection period.

IV. Place all urine tested and collected in bottle #4.

URINE SPECIMEN

30 minutes before bedtime snack test this specimen. Include in container for fourth collection period. Indicate time started. Collect all urine that you pass into bottle #4.

EAT BEDTIME SNACK

URINE SPECIMEN

60 minutes before breakfast collect specimen. Indicate time ended. This ends the **FOURTH** collection for the 24-hour period.

PRESERVATIVES

Many specimen collection procedures need a certain chemical preservative to be aliquoted into the container before urine collection. The specific preservative used depends on the substances analyzed in the specimen. These chemicals are added to preserve the integrity of the specimen. Labels can be made that not only give instructions to the patient but also give the specific preservative and the list of tests that can be done under the described conditions. This helps to ensure that the patient is given the proper container and that the specimen collected will be of value for the test (*see* Fig. 5-1).

Preservatives minimize oxidation and bacterial growth. A change in *p*H can occur as a result of bacterial growth. Acids are used to keep the urine in stable acidic condition (*e.g.*, acetic acid). Because porphyrins are stable in alkaline urine, 5 g of sodium carbonate should be used.

Organic solvents such as toluene and petroleum are used to protect the specimen against oxygen.

Twenty-four-hour collections for the following tests must be made in con-

Fig. 5-1. *Instructions for urine-specimen collection.*

- ☐ Calcium
- ☐ Chloride
- ☐ Creatinine
- ☐ Electrophoresis
- ☐ Iodine
- ☐ Magnesium
- ☐ Murmidase
- ☐ Nitrogen
- ☐ Osmolality
- ☐ Oxalate
- ☐ Phosphorus
- ☐ Protein
- ☐ NaK
- ☐ Thiazide
- ☐ Myeloma
- ☐ Melanin
- ☐ Special Protein
- ☐ ______________

INSTRUCTIONS FOR COLLECTING URINE SPECIMEN —PLAIN MAY INTERCHANGE WITH ACETIC ACID OR TOLUENE—

1. Empty bladder at__________ A.M./P.M.

on__________

THIS URINE **SHOULD NOT** BE SAVED IN THIS BOTTLE: IT **SHOULD** BE **DISCARDED.**

2. Save in this bottle all urine passed from that time until

__________ A.M./P.M. on

____

DAY AND DATE

BLADDER MUST BE EMPTIED INTO THIS BOTTLE AT THE TIME OF THE LAST COLLECTION AS LISTED ABOVE. FLUID INTAKE SHOULD BE WATCHED DURING COLLECTION PERIOD TO INSURE ALL URINE PASSED DOES NOT EXCEED THE CAPACITY OF THIS BOTTLE.

MC 826-03/R477

tainers with the named preservative. There can be no interchange of chemicals listed.

1. *Acetic acid*
   Aldosterone
   Catecholamine
   Cortisol
   Deoxycorticosteroids
   Estriol
   Estrogen
   5-hydroxyindole-acetic acid
   Metanephrine
   Myeloma
   Pituitary
   Pregnanetriol
   Steroids and ketosteroids
   17-keto-steroid fractionation
   Vanillylmandelic acid–homovanillic acid
2. *Hydrochloric acid*
   Histamine
3. *Petroleum ether and sodium carbonate*
   Urobilinogen
4. *Sodium carbonate or petroleum ether*
   Porphyrins
5. *Toluene*
   Amino acids
   Citrate
   Creatine (not creatinine)
   Cyclic AMP
   Cystine
   Fractionated glucose
   Glucose
   Hydroxyproline
   Mucopolysaccharides
   Urea
   Uric acid
   Xanthine

Certain tests require that no preservative be added to the collection container because they may interfere in the test analysis. These tests include chorionic, drugs, and histamine.

Numerous tests are not affected by the addition of a preservative to the collection. The following tests may be done from a collection containing either toluene or acetic acid.

Amylase
Arylsulfatase
Calcium
Creatinine
Chloride
Creatinine clearance
Electrophoresis
Iodine
Magnesium
Melanin

Metals
Muramidase
Nitrogen
Osmolality (plain or toluene)
Oxalate
Phosphorus
Protein
Sodium and potassium
Special proteins

VOLUME OF URINE

An excellent method of determining the urine volume collected is to equate it to the weight of the urine. The empty collection bottle is weighted, and this figure is printed on the bottle itself with indelible pen. After the specimen is collected and returned to the processing area, the bottle with contents is reweighed (Fig. 5-2). By equating the density of the urine to the weight, we know that the weight is equal to the volume of urine. This is a much simpler procedure than using volume itself by measuring in a graduate cylinder.

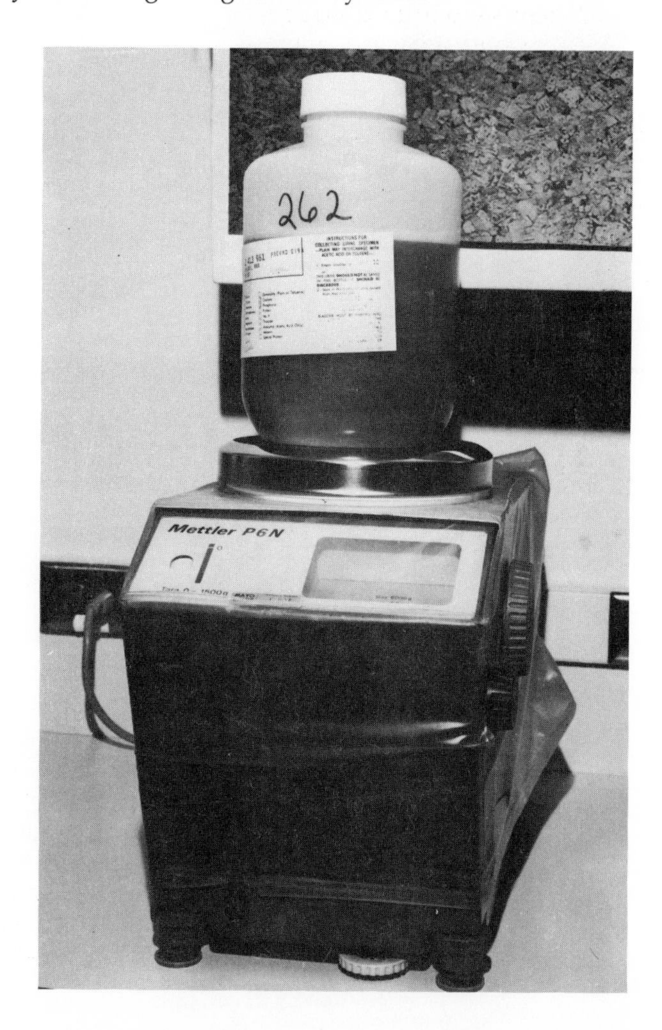

Fig. 5-2. *Urine specimen being weighed. Note bottle weight figure.*

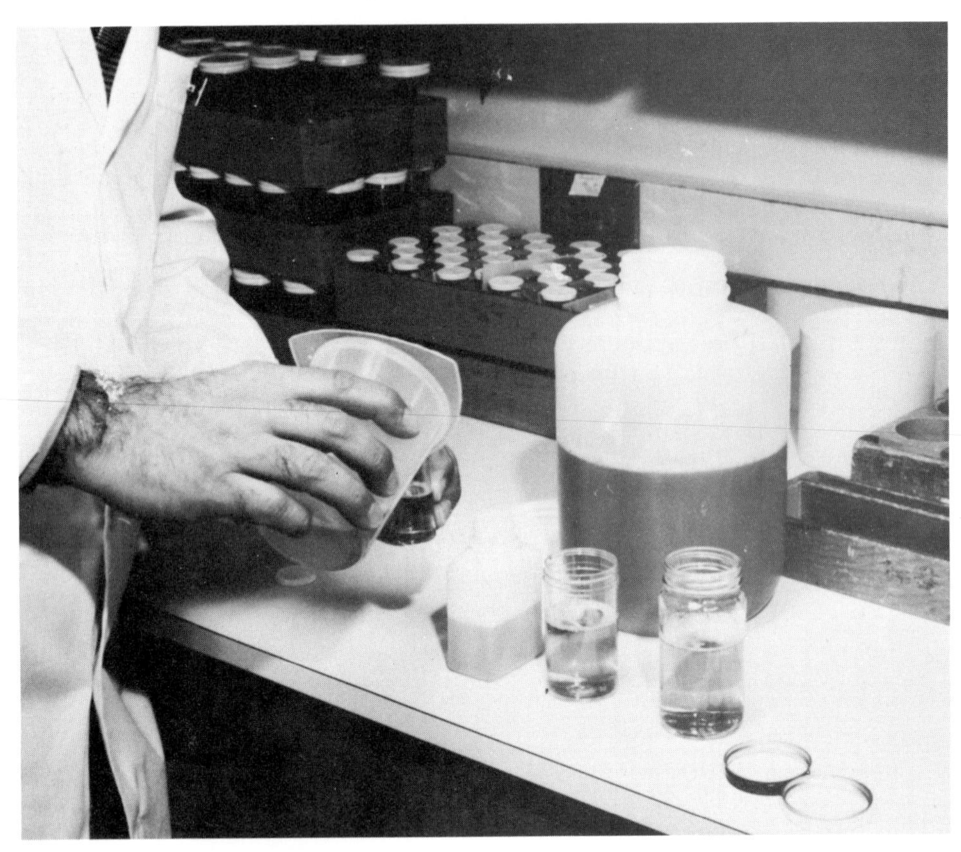

Fig. 5-3. *Aliquoting of timed urine specimens using a disposable plastic measuring cup.*

Fig. 5-4. *Aliquot containers for timed urine specimens.*

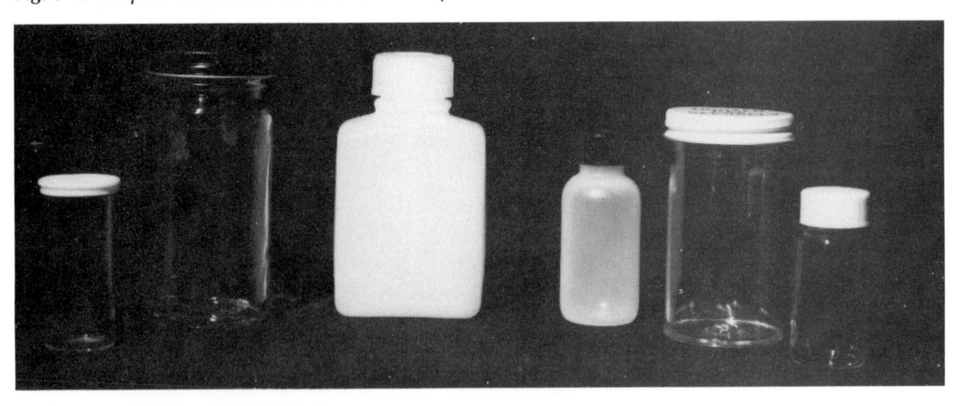

ALIQUOTING

After the urine volume is recorded on each test card, the urine is aliquoted according to the tests ordered (Fig. 5-3). The type of container to be used for the aliquot should be considered carefully, depending on the preservative and tests—for example, if the solvent used is toluene or petroleum, it is better to place the urine aliquot in a glass container than a plastic one. For metals, the urine aliquot should be placed in a special vial that has been found to be metal free. The aliquoting of urine is done in different vials (Fig. 5-4) according to the volume the laboratories requested.

SUMMARY

Quality assurance in urinalysis requires the establishment of certain rules that must be adhered to if good quality results are to be obtained. There must be assurance during specimen collection that clean appropriate containers are being used and that the applicable collection technique is being used. Additionally, there should be an adequate volume of specimen, there should be no visible contamination of the specimen, and the identification and time of collection must be recorded on the request form and container.

REFERENCES

1. Elin RJ: Review of Collection of Urine Specimens for Culture. Bethesda, National Institute of Health, 1979
2. Free AH, Free AM: Urinalysis in Clinical Laboratory Practice. Cleveland, Chemical Rubber Company, 1975
3. Kaas EJ: Asymptomatic infections of the urinary tract. Trans Assoc Am Physicians 69:56–64, 1956
4. Kaas EH: Bacteria and the diagnosis of infection of the urinary tract. Infections Intern Med 100:709, 1957
5. Lennette EH, Spaulding EH, Truant JP (eds): Manual of Clinical Microbiology. Washington DC, American Society of Microbiology, 1974

6

Body Fluid Specimens

Robert I. Kalish
Howard S. Cheskin
Thomas A. Blumenfeld

Pleural, Pericardial, and Peritoneal Fluids

The pleural, pericardial, and peritoneal cavities normally contain small amounts (20–50 ml, or less) of sterile, essentially acellular, serous fluids, formed by plasma ultrafiltration, that serve to lubricate the opposing parietal and visceral surfaces. Abnormal accumulations of fluid in these virtual cavities are referred to as effusions and occur in various diseases, both primary and secondary to the organs they surround. Laboratory examination of effusions can often give helpful diagnostic information about the abnormalities causing their formation and can be a guide to further therapy.

ABNORMAL ACCUMULATION OF FLUIDS IN BODY CAVITIES

Removal of an effusion from a serous body cavity by percutaneous (placing a needle through the skin) aspiration is referred to as *paracentesis*. The specific terms thoracentesis (or thoracocentesis), pericardiocentesis, and peritoneocentesis refer to the pleural, pericardial, and peritoneal cavities, respectively. Fluid from the peritoneal cavity commonly is referred to as ascitic fluid. Similarly, the terms hemothorax, hemopericardium, and hemoperitoneum refer to effusions of blood (usually resulting from gross hemorrhage) within the respective body cavities. Although the term *empyema* often is used specifically to indicate pus in the pleural cavity (pyothorax), it literally means "pus in the body cavity." Unless reference is made to a specific body cavity, the following discussion refers equally to the pleural, pericardial, and peritoneal cavities.

Although the diagnosis of an effusion usually is made by physical examination or radiologic studies, laboratory analysis of the exudative fluid itself usually is needed to determine its cause. In addition to its diagnostic value, paracentesis, which must be performed aseptically by a physician skilled in the procedure and aware of its indications and potential complications,[1] may relieve the symptoms of the effusion, provide life-saving organ decompression, or allow the instillation of drugs in a body cavity.

Because most effusions are large (generally more than 100 ml), obtaining enough fluid for the laboratory to make all necesssary diagnostic tests on a single sample, and thus avoid repeating the procedure, is generally easy, although repeated paracenteses may be needed to alleviate the symptoms and problem of reaccumulated effusions and, rarely, ones that are loculated. Although removal of all fluid in a body cavity may be necessary, many effusions are relatively benign, and serious complications may result if the fluid is removed too rapidly.

EXAMINATION OF FLUIDS FROM BODY CAVITIES

Unless special chemical tests or extensive microbiologic or cytologic examination of the fluid is anticipated, a sample of 50 ml of exudate should be enough for a thorough laboratory evaluation. The paracentesis fluid should be brought to the laboratory promptly (within 1 hour of collection) or, if delay is unavoidable, should be refrigerated at 4°C to avoid bacterial overgrowth, damage to cellular constituents, and alterations in chemical composition. To avoid such factitious results, for most chemical tests a sample of the paracentesis fluid should be allowed to clot, the clot and any other suspended material (*e.g.*, cells) removed by centrifugation, and the supernatant submitted with a specimen of simultaneously drawn serum, which will provide an "internal" reference value for the chemical analysis. The serum should be handled the same way as the effusion fluid.

Routine laboratory evaluation of effusions of unknown cause should include chemical tests (total protein, LDH), microscopic examination (white blood cell count and differential, red blood cell count, cytology), and microbiology examination (Gram stain, acid-fast stain of pleural fluid, aerobic and anaerobic cultures).

EXAMINATION OF COLOR, ODOR, AND VISCOSITY

In addition to chemical analysis of effusions, the gross observations of color, odor, and viscosity may be diagnostically significant, and these should be made when the specimen is obtained and then recorded on both the patient's chart and the laboratory requisition form.

BLOODY FLUID

Serous fluids normally are pale yellow or amber, with a viscosity similar to serum. They lack fibrinogen and do not clot. An effusion containing blood will appear pink when the red blood cell count reaches 5,000 to 10,000 cells/μl and will appear grossly bloody at a red blood cell count of about 100,000/μl.[2] Although the terms serosanguineous and sanguineous sometimes are used to distinguish between serous effusions mixed with blood and effusions primarily composed of blood, these qualitative terms are much less meaningful and less clinically useful than an actual quantitative cell count. The terms hemothorax, hemopericardium, and hemoperitoneum should be reserved for gross intracavitary hemorrhage.

If the effusion contains blood, it is first necessary to determine whether the blood indicates a traumatic tap (*i.e.*, bleeding induced by the paracentesis) or a hemorrhagic effusion reflecting either underlying disease (*e.g.*, malignancy on a serous surface) or whole blood in a body cavity from a ruptured vessel. Nonhomogeneous distribution of blood in the paracentesis fluid, often with clearing as more fluid is removed (confirmable by sequential microhematocrits), is good evidence of a traumatic paracentesis. In addition, bloody fluid from a traumatic tap or active bleeding into the cavity will clot spontaneously if it is collected in a container that does not contain an anticoagulant, whereas serous fluid containing blood that has been in a body cavity for at least several hours will be defibrinogenated and will not clot. Differentiation between a bloody effusion and gross hemorrhage (whole blood) can be accomplished rapidly by comparison of the hematocrit of the freshly drawn paracentesis fluid to that of a simultaneously drawn venous blood specimen; similarity of the two hematocrits reflects active or extremely recent bleeding into the cavity. A bloody, extremely viscous pleural effusion may indicate a mesothelioma (tumor of the lining cells of a body cavity),[3] but such a diagnosis must be based on cytologic findings or tissue biopsy.

Because detection of a hemoperitoneum by abdominal paracentesis may be negative in a high percentage of cases with significant intraperitoneal injury, peritoneal lavage, a more sensitive technique, has been recommended for evaluating blunt abdominal trauma.[4–6] If peritoneal lavage is used, the amount of blood in both the lavage fluid and the prelavage aspirate should be measured because the amount of blood in the peritoneum correlates with the clinical significance of the intraperitoneal injury causing the bleeding.

More than 20 ml of blood aspirated prior to lavage indicates that there is significant intraperitoneal injury. After the aspirate is evaluated, the lavage fluid may be evaluated rapidly for hemoperitoneum and significant intraperitoneal injury by attempting to read newsprint through the fluid.[4] The incidence of significant injury was found to be 100% when newsprint could not be seen through the lavage fluid and at least 94% when the newsprint could be seen but not read. The test was found to be equivocal when the lavage fluid was bloody but newsprint could be read, and a bloodless (clear) fluid correlated with a zero risk of clinically significant injury. A more quantitative assessment of hemoperitoneum and its correlation with abdominal injury can be obtained by red blood cell counts of the lavage fluid; an 85% chance of significant visceral injury was found[5] when the red blood cell count exceeded 100,000/μl (about 22 ml of blood per liter of lavage fluid). Because of possible debris, bacteria, and macrophages in the lavage fluid, the laboratory should make cell counts manually using a hemocytometer rather than with an automated counter. This will avoid both a falsely elevated count and obstruction of the orifice of the automated counter.

A more rapid, semiquantitative measurement of a hemoperitoneum is a hematocrit of the lavage fluid; hematocrits of 2% or more may be indicative of serious injury (20 ml of blood per liter of lavage fluid is equivalent to a hematocrit of about 1%). The clinical usefulness of a white blood cell count and an amylase level of the lavage fluid is questionable.[4]

BILE, URINE, AND FECES IN FLUID

Green peritoneal fluid, obtained by paracentesis or lavage, usually indicates the presence of bile, and a bedside test for bile (*e.g.*, with Labstix) should be performed on all peritoneal aspirates or lavage fluid regardless of the color because bile in the peritoneal fluid is a sign of significant peritoneal injury. The presence of feces or urine in the peritoneal fluid is also of major clinical import; creatinine and urea nitrogen or ammonia assays should be performed if urine is suspected.

CLOUDY OR TURBID FLUID

Cloudy or turbid paracentesis fluid usually indicates leukocytosis (a large number of white blood cells). A high white blood cell count is clinically significant because it may result from various inflammatory and noninflammatory causes. The turbid fluid of leukocytosis, however, may be confused either with the milky white fluid of a chylous effusion (the presence of lymph indicates abnormal lymphatic drainage) or with the milky-to-greenish ("gold paint") appearance of a pseudochlyous effusion.

Rapid differentiation between leukocytosis and a chylous or pseudochylous effusion can be made by centrifugation of the effusion, because turbidity resulting from suspended white blood cells will clear with centrifugation and that from a chylous or pseudochylous will not. Further differentiation between a true chylous and a pseudochylous effusion can be made by acidification of the effusion sample with dilute HCl followed by extraction of the resultant solution with diethyl ether. In a chylous effusion, rich in serum triglycerides and chylomicrons, there will be visible clearing and decreased volume of the lower aqueous phase owing to extraction of the lipid elements into the upper ether layer, whereas the cholesterol-rich pseudochylous effusion should neither clear nor decrease in volume. Final differentiation between the two types of effusion usually may be made by microscopic examination and quantitative lipid analysis of the fluid, preferably of a specimen obtained from the patient after a fast.[1, 7, 8]

CHEMICAL TESTS

Analysis of the protein and LDH in effusions is particularly important if a distinction is to be made between a transudate and an exudate. A transudate is a protein-poor fluid caused by hydrostatic factors (extrinsic to the surfaces of the body cavity) that influence the rates of formation or resorption of serous fluid. An exudate is a protein-rich fluid caused by diseases of the organs in, or surfaces of, the body cavity, with resultant damage to the mesothelial cell lining the body cavities.[9] This differention should be the first step in the evaluation of a pleural effusion.[9] The differentiation of peritoneal and pericardial exudates from transudates by laboratory analysis is less well established,[10] although almost all pericardial effusions are exudates.

Other studies of paracentesis fluid that may be of particular diagnostic value in specific clinical circumstances but which generally are not routinely performed are glucose; amylase; *p*H; ammonia; creatinine and urea nitrogen; alkaline phosphatase; lipids; complement; and hyaluronate.

If *p*H determination of an effusion is requested, the sample should be obtained anaerobically in a heparinized glass syringe (the same as for an arterial blood gas sample, which simultaneously should be obtained for reference purposes), maintained at 0°C (ice bath) under anaerobic conditions (tip of needle or syringe sealed), and analyzed using a *p*H electrode within 20 minutes of collection, if possible. Both the *p*H and P_{CO_2} of pleural fluid are stable under such conditions.[11] The clinically important recognition of markedly acidic (*p*H less than 6) pleural effusions associated with perforation of the esophagus can be made rapidly and simply by the use of *p*H reagent paper.[12]

If pleural fluid[13] or pericardial fluid[14] complement assays are requested (*e.g.*, in suspected cases of rheumatoid arthritis or lupus erythematosus), the effusion should be collected either with or without EDTA anticoagulant (depending on the specific assay requested), centrifuged, frozen within 2 hours of collection, and stored at −70°C until the desired assays are performed; plasma or serum samples must be obtained simultaneously for reference and stored at −70°C.[14]

MICROBIOLOGIC STUDIES OF BODY CAVITY FLUIDS

Although microbiologic evaluation of normally sterile paracentesis fluids may be recommended only when an infectious cause is strongly suspected (*e.g.*, an aspirate with pus or a foul odor),[10, 15] it is preferable to include microbiology routinely as part of the evaluation of an effusion.

Bacteria

A specimen for general aerobic bacteriology should contain 3 ml to 5 ml of fluid and should be collected in a screw-capped sterile container containing a few drops of sterile sodium heparin (without preservatives) or EDTA to prevent the formation of clots that may trap any microorganisms (or cells) present. Although all such microbiologic specimens must be processed as quickly as possible, specimens of pericardial fluid should have immediate priority because of the importance of a diagnosis of bacterial pericarditis.

After the appropriate samples of fluid have been removed for culture and sent to the laboratory, a Gram stain should be prepared directly from a few drops of the aspirated fluid (if purulent) or from the sediment obtained after centrifugation (2500 rpm, 30 minutes) of the fluid to maximize the chance of observing any bacteria that may be present. Fluid from peritoneal lavage has already been diluted and should always be centrifuged before a Gram stain. The results of the Gram stain should be noted on the requisition form because it may be helpful in the proper laboratory processing of the specimen, especially if "unusual" forms such as spores, branching Gram-positive elements, or sulfur granules are seen. The requisition form must also include a record of all recent antibiotic therapy because such previous therapy may result in false-negative bacterial cultures. Any recent invasive procedures should also be noted because these may have contaminated the body cavity. In addition to the sample of paracentesis fluid submitted, blood cultures (aerobic and anaerobic) and other appropriate samples (*e.g.*, pleural, pericardial, or peritoneal biopsy samples, wound aspirates) should also be submitted because they may provide diagnostic guidance if the effusion cultures are negative. The same care must be taken in obtaining and processing these additional specimens as in the handling of the original effusion samples.

In addition to the Gram stain, other microbiologic tests that may be done on paracentesis fluid include capsular swelling (Quellung) tests with specific antisera for *Streptococcus pneumoniae, Hemophilus influenzae,* and *Neisseria meningitidis;*[17] the detection of microbial antigens or antibodies by immunoprecipitation techniques;[17, 18] and a *Limulus* amoebocyte lysate assay for the endotoxin of gram-negative organisms;[19, 20] if the *Limulus* lysate assay is used, endotoxin-free glassware and sterile disposable plastic equipment are preferred.

If peritonitis is clinically suspected as a result either of trauma or disease, a white blood cell count (including a differential count) should be performed immediately after incubation of the bacterial cultures and after evaluation of the Gram stain because, even without a positive Gram-stain or culture, it may be best to begin antibiotic therapy whenever the peritoneal fluid contains more than 500 white blood cells/μl with more than 50% polymorphonuclear neutrophils.[21]

Infection of the three major body cavities by *Neisseria* species is uncommon but can occur, and if gram-negative intracellular or extracellular diplococci are seen or if neisserial infection is suspected clinically, plates of modified Thayer–Martin or New York City culture media should be inoculated at the time of the paracentesis, immediately incubated at 35°C to 37°C in a humid atmosphere of 3% to 10% CO_2 (candlejar), and examined by the laboratory after 24 to 48 hours of incubation. If the effusion cannot be processed in this ideal manner, a nutritive transport medium such as the "Transgrow" bottle system or the JEMBEC plate should be inoculated at the bedside and incubated for 18 to 24 hours before delivery to the laboratory. Like *Neisseria, Hemophilus influenzae* is a fragile organism, but direct bedside inoculation of Thayer–Martin chocolate agar wth incubation at 35°C to 37°C in a moist CO_2 atmosphere should ensure survival of the organism until the samples reach the laboratory.

Mycobacteria

Relatively few organisms may be shed into a body fluid in a mycobacterial infection of an enclosed organ, and relatively large amounts of an effusion (at least 5–10 ml) should be submitted in a separate tube. Even with a sufficient sample, only 25% to 30% of tubercular pleural effusions may yield a positive *Mycobacterium tuberculosis* culture.[22, 23] The percentage of positive cultures can, however, be increased by culture of the sediment obtained from centrifugation of large (100–500 ml) volumes of pleural fluid, by performance of multiple cultures, and by guinea pig inoculations.[23] Culture of a pleural biopsy specimen may be positive in up to 80% of cases of tuberculous pleurisy, and, if possible, a biopsy sample should be submitted in sterile saline for culture with the pleural fluid sample if a diagnosis of tuberculous pleural effusion is strongly suspected. An even higher percentage of positive results (about 95%) may be obtained if the pleural biopsy is cultured and examined histologically with an acid-fast stain.[24] Some authors recommend that pleural fluid culture, pleural biopsy culture, and pleural biopsy histology be performed on all patients with pleural effusions because only one of these examinations may be positive for mycobacteria.[23]

Samples for mycobacteriologic study should be kept at 4°C until brought to the laboratory to prevent overgrowth by exogenous or endogenous organisms. If delays are anticipated in delivery of the samples to the laboratory, the paracentesis fluid may be inoculated into a noninhibitory culture medium such as Middlebrook 7H9 broth to enhance mycobacterial growth and to aid in laboratory identification of the organism. In general, selective (inhibitory) culture media containing antibiotics should not be used because overgrowth of the medium by coexisting bacteria or fungi usually is not a problem with normally sterile body fluids in the absence of polymicrobial infection.

All personnel working with paracentesis fluid and the paracentesis equipment should wear a surgical gown, mask, and gloves; avoid direct contact with the fluid; and should prevent aerosol formation, which can spread droplets that contain mycobacteria. All specimens of paracentesis fluid suspected of containing mycobacteria should be processed only in a properly equipped laboratory. The only procedure that may safely be performed outside such a laboratory is the preparation of an acid-fast stain of the sediment obtained from a centrifuged sample of the effusion.[25] Because of the small number of mycobacteria typically present in a body fluid, an acid-fast stain may not demonstrate any mycobacteria, but since any acid-fast organism seen is of pathologic significance, great care must be taken in preparing and examining acid-fast stains; use of a fast and sensitive fluorochrome stain such as auramine-rhodamine is recommended. Even if acid-fast organisms are seen, a culture is still required because infection may be caused by species other than *M. tuberculosis*. If the patient has received antimycobacterial drug therapy, it is important that this information be written on the laboratory requisition form.

Anaerobic Bacteria

Techniques for the proper collection, processing, and laboratory evaluation of anaerobic organisms have become more refined, and the number of these bacteria discovered in infections of the three major body cavities has increased, as evidenced by an 80% anticipated yield from peritoneal fluid in peritonitis; 50% to 75% expected yields from pleural fluid in empyema; and 85% expected yield from pleural fluid in lung abscesses. All paracentesis fluid should be submitted routinely for anaerobic and aerobic cultures.[26], [27] Although a strong, foul odor, when present, is characteristic of anaerobic infections, the absence of such an odor does not exclude an anaerobic or mixed aerobic and anaerobic infection.[28]

The aspiration of pus from a body cavity establishes the diagnosis of empyema and is an exceedingly important finding. Empyema often may be loculated and require multiple paracenteses to obtain purulent fluid. Fluid from all such procedures must be submitted for aerobic and anaerobic cultures; in one study of 83 cases of empyema, anaerobes were recovered in 76% of the cases and were the only isolates in 35% of the cases.[29] Empyema due to mycobacteria is relatively uncommon today.

Although anaerobic culture media sometimes should be inoculated at the bedside, such procedures generally are impractical and probably unnecessary provided that the sample is properly collected, anaerobically stored, and quickly transported to the laboratory. If the anticipated transport time to the laboratory is about 30 minutes or less, the specimen may be submitted in a syringe from which all air has been expelled and with the needle stuck into a sterile stopper. Such a specimen is also suitable for general bacteriologic cultures. If longer transport periods are involved, the sample should be submitted in a degassed collection tube containing a preduced, anaerobically sterilized (PRAS) transport medium such as the commercially available Carey–Blair or Amies media. Alternatively, the paracentesis fluid may be inoculated into either a commercially available tube containing oxygen-free CO_2 or an anaerobic transport tube such as the commercially available Port-a-cul tube or placed in a chamber made anaerobic by a GasPak generator.

Fungi

Only small numbers of fungi usually are seeded from infected tissue into body fluids, and relatively large amounts (at least 5–10 ml) of effusion should be submitted, preferably in a separate tube, for mycologic studies.

Most systemic fungal pathogens grow slowly, and examination of effusions for fungi should start at the bedside with a direct microscopy of an unstained wet mount or stained smear of fresh body fluid, preferably of the sediment formed during centrifugation ($1000 \times g$, 15 minutes). This is particularly true in suspected cases of *Cryptococcus neoformans* infection where examination of a smear of centrifuged sediment containing India ink or nigrosin as a negative capsule stain may allow rapid detection of the encapsulated organism.[30] This yeast may also be detected rapidly by microscopic examination of dry smears of the centrifuged sediment that have been stained with alcian blue or mucarmine.[31]

Body fluids suspected of containing fungi should be inoculated into a culture medium at the bedside within 1 hour of collection. For this purpose, Sabouraud dextrose agar slants in screw-capped tubes or bottles are preferred, although Sabouraud agar containing the antibiotics chloramphenicol and cycloheximide (Mycosel agar) is useful in preventing bacterial overgrowth if prolonged storage or transport of the sample to an outside mycology laboratory is anticipated. If media containing antibiotics are used, it is important to realize that these antibiotics are inhibitory to the growth of a number of fungi and should not be the sole medium used for their isolation. If these media are used, a reference sample of uninoculated fluid should be retained and kept at 4°C.[32]

If delays in the processing of uninoculated samples of paracentesis fluids are unavoidable, the samples should be kept at 30°C to accelerate the rate of fungal growth and shorten the time required for identification once the specimens reach the laboratory. Bacterial overgrowth in samples may be prevented by the addition of a combination of antibiotics such as penicillin (20 units/ml of fluid) and streptomycin (40 units/ml of fluid), although antibiotics must not be added to effusions suspected of containing the "fungal-like" bacteria *Nocardia* species or *Actinomyces* species.

Parasites

Parasitic disease is a rare cause of effusions, but infections by parasites such as *Echinococcus granulosus*,[33] *Mansonella ozzardi*,[34] and *Entamoeba histolytica*[35] have been diagnosed by identification of the organism in appropriately stained specimens of pleural fluid, ascitic fluid, and pericardial fluid, respectively. In general, if parasitic disease is suspected as a cause of an effusion, a parasitologist should be contacted, preferably before the paracentesis. The sample should be obtained aseptically, placed in a tightly sealed screw-capped container, kept at ambient temperature (24–28°C), and brought quickly to the parasitology laboratory. Under no circumstances should samples suspected of containing parasites be heated or frozen.[36] Anchovylike material in an effusion is often thought to be classic evidence of an amoebic abscess, but this material is found only occasionally.[37]

Viruses

Viruses may be a cause of cavitary effusions,[2] and, if suspected, a virologist should be consulted before the paracentesis is performed and all appropriate precautions (gowns, surgical masks, and gloves) taken during the collection and handling of the effusion and the disposal of the paracentesis equipment used.

If possible, at least 3 ml to 5 ml of the body fluid should be placed in a dry, sterile, screw-capped glass container that does not have a virus transport medium. Because many pathogenic viruses are quite labile at room temperature, rapid transportation of the specimen to the laboratory is necessary; if possible, the appropriate tissue culture system should be inoculated at the bedside. Since freezing can destroy the infectivity of some pathogenic viruses, the specimen of body fluid should be transported at ambient temperature or, if a delay of not more than a few hours is anticipated, kept at 4°C (melting ice). Specimens that require long transport times should be quickly ("snap") frozen at −70°C or lower (Dry Ice, Dry Ice-alcohol, liquid nitrogen) and shipped at this temperature. Fluctuations in temperature with concomitant freezing and thawing of the sample must be avoided. Because shipment under such rigidly controlled conditions is often not feasible, it is best to ship, in an insulated container, these samples unfrozen at 4°C in a tightly sealed, screw-capped glass vial with the cap covered with adhesive tape. Provision should be made to keep the sample at 4°C (and not frozen) if delays in transit are expected. Detailed instructions for the shipment of such specimens are given in a World Health Organization monograph.[38]

Generally, the shedding of viruses into body fluids usually decreases rapidly after the onset of illness, and specimens of body fluids for diagnostic virology should be collected as soon as possible after clinical onset of disease (acute-phase specimen). In addition, although the etiologic virus may be isolated from a body fluid, it may also be isolated from, or detected in, other clinical specimens such as throat swabs, vesicle fluid, urine, stool, and tissue. This material must be collected properly and submitted with the specimen of body fluid.[39] Because of the frequent use of serologic tests in the diagnosis of viral disease, 5 ml to 10 ml of serum should be collected during the acute phase of the viral illness and kept frozen (−20– −70°C) until needed.

Since viruses require living cells for replication, viral titers decline rapidly after death, and postmortem samples of body fluids for virologic study must be collected aseptically as quickly as possible after death. Because bacterial overgrowth, which will interfere with viral isolation, may occur rapidly after death, it is probably advisable to add antibiotics to such specimens (or to any other virologic specimen in which contamination by exogenous or endogenous bacteria is possible); a satisfactory combination consists of penicillin (500 units/ml), gentamicin (50 μg/ml), and amphotericin B (10 μg/ml).

CYTOLOGY

Examinations of cells in effusions may be of great value in the diagnosis of cancer or nonmalignant conditions.[40] As much fluid as possible up to a maximum of about 200 ml should be collected for cytologic study in either a tube or syringe. Sodium heparin (5–10 units/ml) or EDTA (1 mg/ml) may be added to the fluid as an anticoagulant, but sodium citrate and oxalate should not be used.[41] Ideally, specimens of pleural, peritoneal, or pericardial fluid should be examined unfixed, as soon as possible after collection, to minimize changes in cell structure. Cells in these body fluids are relatively well preserved for 24 to 48 hours if the specimens are kept refrigerated.[42] If these effusions cannot be examined within 48 hours of collection, they should be prefixed within 1 hour of collection in an equal volume of 50% ethanol (added to the fluid). The cellular morphology of body fluid treated in this way is preserved for at least 1 week. Prefixation, which coagulates proteins and hardens cells into spherical shapes, should be avoided if membrane filters are used for cellular concentration. Either fresh or prefixed specimens are, however, suitable for preparing direct or sediment smears on glass slides.

STARCH

Starch peritonitis may occur 10 days to 30 days after intraabdominal surgery and may mimic bacterial peritonitis. This condition may be diagnosed by immediate microscopic identification of starch granules in the amber ascitic fluid using polarized light microscopy or iodine staining.[43] A sample of the ascitic fluid should be submitted for routine aerobic and anaerobic cultures, a white blood cell count, and a differential count to ensure that a postoperative bacterial infection is not present.

REFERENCES

1. Krieg A: Cerebrospinal fluid and other body fluids. In Henry JB (ed): Todd–Sanford–Davidsohn Clinical Diagnosis and Management by Laboratory Methods, 16th ed, pp 667–674. Philadelphia, WB Saunders, 1979
2. Ayvazian LF: Diagnostic aspects of pleural effusion. Bull NY Acad Med 53:532–536, 1977
3. Hellström P–E, Friman C, Teppo L: Malignant mesothelioma of 17 years duration with high pleural fluid concentration of hyaluronate. Scand J Respir Dis 58:97–102, 1977
4. Jergens ME: Peritoneal lavage. Am J Surg 133:365–369, 1977
5. Engrav LH, Benjamin CI, Strate RG, Perry JF Jr: Diagnostic peritoneal lavage in blunt abdominal trauma. J Trauma 15:854–859, 1975
6. Olsen WR, Redman HC, Hildreth DH: Quantitative peritoneal lavage in blunt abdominal trauma. Arch Surg 104:536–543, 1972
7. Seriff NS, Cohen ML, Paul S, Schulster PL: Chylothorax: Diagnosis by lipoprotein electrophoresis of serum and pleural fluid. Thorax 32:98–100, 1977

8. Lesser GT, Bruno MS, Enselberg K: Chylous ascites: Newer insights and many remaining enigmas. Arch Intern Med 125:1073–1077, 1970
9. Light RW, MacGregor I, Luchsinger PC, Ball WC Jr: Pleural effusions: The diagnostic separation of transudates and exudates. Ann Intern Med 77:507–513, 1972
10. Glasser L: Body fluid evaluation: serous fluids. Diagn Med Sept/Oct:79–90, 1980
11. Potts DE, Levin DC, Sahn SA: Pleural fluid pH in parapneumonic effusions. Chest 70:328–331, 1976
12. Dye RA, Laforet EG: Esophageal rupture: diagnosis by pleural fluid pH. Chest 66:454–455, 1974
13. Hunder GG, McDuffie FC, Huston KA, Elveback LR, Hepper NG: Pleural fluid complement, complement conversion, and immune complexes in immunological and nonimmunological diseases. J Lab Clin Med 90:971–980, 1977
14. Hunder GG, Mullen BJ, McDuffie FC: Complement in pericardial fluid of lupus erythematosus: Studies in two patients. Ann Intern Med 80:453–458, 1974
15. Storey DD, Dines DE, Coles DT: Pleural fluid: A diagnostic dilemma. JAMA 236:2183–2186, 1976
16. Austrian R: The Quellung reaction, a neglected microbiological technique. Mt Sinai J Med (NY) 43:699–709, 1976
17. Coonrod JD, Rytel MW: Detection of type-specific pneumococcal antigens by counterimmunoelectrophoresis: I. Methodology and immunologic properties of pneumoccocal antigens. J Lab Clin Med 84:770–777, 1973
18. Coonrod JD, Rytel MW: Detection of type-specific pneumoccocal antigens by counterimmunoelectrophoresis: II. Etiologic diagnosis of pneumoccocal pneumonia. J Lab Clin Med 81:778–786, 1973
19. Nachum R. Neely M: Clinical diagnostic usefulness of the Limulus amoebocyte lysate assay. Lab Med 13:112–117, 1982
20. Rytel MW: Rapid diagnostic methods in infectious diseases. Adv Intern Med 20:37–60, 1976
21. Conn HO: Spontaneous bacterial peritonitis: Multiple revisitations. Gastroenterology 70:455–457, 1976
22. Falk A: Tuberculous pleurisy with effusion: Diagnosis and results of chemotherapy. Postgrad Med 38:631–635, 1965
23. Berger HW, Mejia E: Tuberculous pleurisy. Chest 63:88–92, 1973
24. Levine H, Metzger W, Lacera D, Kay L: Diagnosis of tuberculous pleurisy by culture of pleural biopsy specimen. Arch Intern Med 126:269–271, 1970
25. Runyon EH, Karlson AG, Kubica GP, Wayne LG (revised by Sommers HM, McClatchy JK): Mycobacterium. In Lennett EH, Ballows A, Hausler WJ Jr, Truant JP (eds): Manual of Clinical Microbiology, 3rd ed, pp 150–179. Washington, D.C., American Society for Microbiology, 1980
26. Allen SD, Sider JA: Procedures for isolation and characterization of anaerobic bacteria. *See* Reference 25, pp 397–417
27. Talley FP, Bartlett JG, Gorback SL: Practical guide to anaerobic bacteriology. Lab Med 9(9):26–35, 1978
28. Bartlett JG, Finegold SM: Anaerobic infections of the lung and pleural space. Am Rev Respir Dis 110:56–77, 1975
29. Bartlett JG, Gorbach SL, Thadepalli H, Finegold SM: Bacteriology of empyema. Lancet 1:338–340, 1974
30. Rinaldi MG: Cryptococcosis. Lab Med 13:11–19, 1982
31. Silva-Hutner M, Cooper BH: Yeasts of medical importance. *See* Reference 25, pp 562–576
32. Larsh HW, Goodman NL: Fungi of systemic mycoses. *See* Reference 25, pp 577–594
33. Jacobson ES: A case of secondary echinococcosis diagnosed by cytological examination of pleural fluid and needle biopsy of pleura. Acta Cytol 17:76–79, 1973
34. Figueroa JMV: Presence of microfilariae of *Mansonella ozzardi* in ascitic fluid. Acta Cytol 17:73–75, 1973
35. Kean BH: Parasitic diseases of the heart. In Gould SE (ed): Pathology of the Heart and Blood Vessels, 3rd ed, pp 812–833. Springfield, Illinois, Charles C Thomas, 1968

36. Miller JH: Clinical parasitology: Introduction and collection of specimens. *See* Reference 25, pp 669–687
37. Kory RC, Smith JR: Laboratory aids in investigating pulmonary diseases. In Baum GL (ed): Textbook of Pulmonary Diseases, 2nd ed, pp 27–83. Boston, Little, Brown, & Co, 1974
38. Madeley CR: Guide to the Collection and Transport of Virological Specimens. Geneva, World Health Organization, 1972
39. Ray CG, Hicks MJ: Laboratory diagnosis of viruses, rickettsia, and chlamydia. *See* Reference 1, pp 1814–1879
40. Koss LG: Diagnostic Cytology and Its Histopathogical Bases, 3rd ed, pp 878–970. Philadelphia, JB Lippincott, 1979
41. Cardozo PL: A critical evaluation of 3,000 cytologic analyses of pleural fluid, ascitic fluid and pericardial fluid. Acta Cytol 10:455–460, 1966
42. Bales CE, Durfee GR: Cytological Techniques. *See* Reference 40, pp 1187–1266
43. Warshaw AL: Diagnosis of starch peritonitis by paracentesis. Lancet 2:1054–1056, 1972

Synovial Fluid

Synovial fluid (SF), a plasma ultrafiltrate that contains hyaluronoprotein, is elaborated by the lining of diarthrodial (synovial) joints, and provides lubrication and nourishment for the joint cartilage. Although normally present in small quantities (less than 3–4 ml in a normal knee joint), the amount of SF can be greatly increased in various joint diseases, and analysis of the SF can be valuable for diagnosing the cause of joint disease.[1–5]

FLUID COLLECTION

Percutaneous aspiration of a joint effusion is called *arthrocentesis* and must be performed under aseptic conditions by a person skilled in the procedure.[6–8] All fluid in the synovial cavity should be aspirated into the collecting syringe, preferably by using a disposable needle and a disposable plastic syringe, which avoids contamination of the aspirated fluid by any birefringent material introduced in washing and resterilizing procedures. Unless an effusion owing to joint disease is present, a "dry" tap, reflective of the small amount of fluid normally present, usually will result. Although there is no absolute "correlation between the volume of an effusion and the cause, severity, or duration of joint disease,"[8] the amount of SF present in an inflamed joint is generally maximal when the inflammation is most severe and tends to decrease with the inflammation; thus the volume of SF removed should be recorded. If there is doubt as to whether the joint space has been entered, a small amount of the aspirated fluid can be expelled from the syringe into a test tube and identified as SF because of increased turbidity or clot formation when 2% acetic acid is added or by its metachromasia when stained with toluidine blue.[9] Even if no SF is visible within the collecting syringe, enough may be present in the syringe needle to provide useful diagnostic information, especially in bacterial or crystal-induced arthritis. The fluid either can be expressed from the needle at the bedside for immediate inoculation of culture media or the needle tip can be placed into a sterile rubber stopper and the syringe and attached needle sent to the laboratory. Synovial fluid should never be sent to the laboratory on cotton swabs. If possible, arthrocentesis should be performed on a patient who has been fasting for at least 4 to 6 hours (preferably overnight) to allow equilibration between SF and blood glucose; a blood sample should be drawn simultaneously and submitted for serum glucose determination.

LABORATORY TESTS

Normal laboratory analysis of SF should include microbiology, microscopic examination (crystals, white blood cell count and differential, other cell studies such as rheumatoid arthritis cells and lupus erythematosus cells), chemistry (glucose, protein, enzymes), and immunologic studies when appropriate (*e.g.*, rheumatoid factor, antinuclear antibodies, immunoglobulins, complement). The volume, gross appearance, viscosity, and ability to clot are valuable diagnostically and can be evaluated at the bedside when the SF is fresh. Often tests are performed at the bedside, the data obtained from these tests can be put on both the patient's chart and the sample requisition form.

FLUID APPEARANCE

The gross appearance of the SF can aid in diagnosis of joint disease. Normal SF is colorless (or very pale yellow), clear, and nonturbid. Turbidity may be caused by a number of factors, including the presence of white blood cells, red blood cells, crystals, and microorganisms; can be assessed by the ability to read newsprint through the tube of SF; and may be graded from 0 (crystal clear fluid) to 4+ (newsprint cannot be read through tube). The SF may appear to be falsely turbid if it is viewed only through a plastic collecting syringe. Coloration of the SF is abnormal and must be noted; in particular, green-colored SF may be observed with *Hemophilus influenzae* septic arthritis,[10] acute crystal-induced arthritis,[8] and chronic rheumatoid arthritis.[1]

If blood is seen in the SF during the arthocentesis, its heterogeneity or homogeneity should be evaluated. Traumatic taps usually give bright red, unevenly distributed blood that may either decrease in amount as the aspiration proceeds or, more commonly, first appear in the syringe toward the end of the arthrocentesis. Blood present in the joint before the tap (hemarthrosis) is usually red-to-brown and homogeneously distributed. A relatively small amount of blood (5–10%) mixed with SF because of a traumatic tap may be grossly indistinguishable from whole blood owing to hemarthrosis. A microhematocrit can be obtained to differentiate between them because the microhematocrit of SF from a traumatic tap is less than that of venous blood, and the microhematocrit from a hemarthrosis is approximately equal to that of venous blood. Also, blood from a traumatic tap usually is not defibrinogenated and may clot during or shortly after the arthrocentesis, whereas blood from hemarthroses, which may be caused by fracture, tumor, or hemophilia, will clot only when the hemarthrosis has immediately preceded or is coincident with the arthrocentesis.

Evaluation of xanthochromia (yellow color) in the supernatant of centrifuged SF may be difficult because normal SF may have a pale yellow color. A dark yellow, orange, pink, or red color in the supernatant after a bloody tap strongly suggests that hemarthrosis has occurred at least 1 to 2 hours before the arthrocentesis; thus an evaluation of xanthochromia should be made within 1 hour of the collection of the SF.

FLUID VISCOSITY

After assessment of the gross appearance of the freshly drawn SF, its viscosity should be evaluated qualitatively by the string test, which measures how far a drop of SF will "string" before breaking when slowly expressed from the collecting syringe after removal of the needle. As an alternative test, a drop of SF can be compressed between the thumb and index finger (which should be gloved to avoid possible contamination of the examiner by infected SF); the thumb and index finger then are pulled apart suddenly to determine how far the fluid "string" will stretch before snapping. Values of less than 3 cm to 6 cm indicate an abnormally low viscosity, usually indicative of a chronic inflammatory joint disease, although an effusion after injury can also result in decreased viscosity. For accurate measurements of viscosity, either an Ostwald viscometer, which requires about 5 ml of SF, or a white blood cell diluting pipette, which requires only about 0.33 (or less) ml of heparinized SF and is more convenient to use, may be employed.[11]

If the SF is so viscous that handling it for subsequent tests is difficult, its viscosity can be reduced to that of serum either by dilution or by incubation with 1 mg (300 NF units) of bovine testicular hyaluronidase per milliliter of SF at 37°C for 30 minutes.[12]

FLUID CLOTTING

The mucin clot (Ropes) test, which reflects the nature of the hyaluronoprotein in the SF, should be performed by the addition of one part of SF to four parts (1–2 ml) of 2% to 5% acetic acid in a small beaker. After about 1 minute, SF from normal joints and from joints with noninflammatory diseases produces a firm, ropy clot which does not break up readily when agitated and which is surrounded by a clear solution (graded as good). In contrast, the SF from joints with active inflammatory disease yields a soft clot surrounded by a slightly turbid solution (graded as fair), a friable clot which breaks up readily when agitated and which is surrounded by a turbid solution (graded poor), or just a cloudy suspension without a true clot (very poor). Alternatively, the test may be performed by addition of a few drops of glacial acetic acid to the surface of the supernatant obtained from a sample of centrifuged SF; the clot, which forms as the heavier acid settles to the bottom of the tube, can be evaluated as before.

Table 6-1. *Gross Analysis of Joint Fluid*

CRITERIA	NORMAL	NONINFLAMMATORY (GROUP I)	INFLAMMATORY (GROUP II)	PURULENT (GROUP III)
1. *volume (ml) (knee)*	<4	Often >4	Often >4	Often >4
2. *Color*	Clear to pale yellow	Xanthochromic	Xanthochromic to white	White
3. *Clarity*	Transparent	Transparent	Translucent to opaque	Opaque
4. *Viscosity*	Very high	High	Low	Very low, may be high with coagulase-positive staphylococcus
5. *Mucin clot**	Good	Fair to good	Fair to poor	Poor
6. *Spontaneous clot*	None	Often	Often	Often

*Recent effusions do not give firm clot because of serum admixture.

CRYSTALS

In cases of suspected crystal-deposition disease (gout, pseudogout), a "wet drop" preparation may be made at the bedside for immediate microscopic examination (preferably using compensated polarized light microscopy), and the results reported on the sample requisition form. If the freshly obtained SF is stored in a refrigerator for several hours before making the "wet drop" preparation, crystals, which were not present initially, may be observed, leading to an erroneous diagnosis of gout.[13] The "wet drop" specimen, which will keep for several hours, is prepared by placing about one drop of freshly obtained SF on a cleaned (alcohol or acetone), unscratched, dust-free (dust is positively birefringent), clear glass slide. A coverslip is then placed over the SF so that the fluid just reaches the coverslip's edge. The edges of the coverslip are then sealed with clear nail polish, which is allowed to dry thoroughly before microscopic examination. (This avoids damage to the objective lens of the microscope.)

After gross analysis of the SF (volume, color, clarity, viscosity, ability to clot, presence of blood) has been completed, the SF may be considered to be normal or may be assigned to one of four pathologic groups, each of which is compatible (with overlap) with a number of disease states (*see* Tables 6-1, 6-2). Thus, careful bedside evaluation of the SF may narrow considerably the differential diagnosis of the arthropathy even before the SF has been sent to the laboratory.

MICROBIOLOGIC EXAMINATION OF SYNOVIAL FLUID

Septic arthritis can result from hematogenous dissemination of an infection to the joint space, local extension from an infected bone or soft tissue, or direct injection of a microorganism into the joint capsule (*e.g.*, trauma, surgery) and

Table 6-2. Diseases That Produce Fluids of Different Groups

NONINFLAMMATORY (GROUP I)	INFLAMMATORY* (GROUP II)	PURULENT* (GROUP III)	HEMORRHAGIC (GROUP IV)
Osteoarthritis	Rheumatoid arthritis	Bacterial infections	Trauma, especially fracture
Early rheumatoid arthritis	Reiter's syndrome	Tuberculosis	Neuroarthropathy (Charcot joint)
Trauma	Crystal synovitis, acute (gout and pseudogout)		Blood dyscrasia (e.g., hemophilia)
Osteochondritis dissecans	Psoriatic arthritis		Tumor, especially pigmented villonodular synovitis or hemangioma
Aseptic necrosis	Arthritis of inflammatory bowel disease		Chondrocalcinosis
Osteochondromatosis	Viral arthritis		Anticoagulant therapy
Crystal synovitis; chronic or subsiding acute (gout and pseudogout)	Rheumatic fever		Joint prostheses
†Systemic lupus erythematosus			Thrombocytosis
†Polyarteritis nodosa			Sickle-cell trait or disease
Scleroderma			
Amyloidosis (articular)			

(McCarty DJ (ed): Arthritis and Allied Conditions: A Textbook of Rheumatology, 9th ed. Philadelphia, Lea & Febiger, 1979)

*As a disease in these groups remits, the exudate (fluid) passes through a group I phase before returning to normal.

† May occasionally be inflammatory, usually when clinical picture is that of rheumatoid arthritis.

can result in rapid, irreversible destruction of a joint. All material from suspected joint infections should have immediate priority because proper treatment may depend on the results of the microbiologic examination of the SF. Microbiologic examination should begin at the bedside with a Gram stain of a smear prepared from a few drops of SF (use of the sediment from centrifuged SF will maximize the chances of observing any bacteria present). The results of the Gram stain must be noted on the requisition form.

BACTERIA

Even if gram-negative intracellular or extracellular diplococci are not seen on Gram stain (about 50% of Gram stains in cases of gonococcal arthritis are negative) and gonococcal arthritis is suspected clinically, plates of the appropriate nutritive culture media for *N. gonorrhoeae* (modified Thayer–Martin or New York City media) should be inoculated at the time of the arthrocentesis, incubated immediately at 35°C to 37°C in a humid atmosphere of 3% to 10% CO_2 (*e.g.*, in a candlejar), and examined by the laboratory after 24 hours to 48 hours of incubation at this temperature. This procedure will maximize the chance of recovering the exceedingly fragile *N. gonorrhoeae,* which may be present only in small numbers in the SF and which may not survive prolonged storage or temperature variation. If the sample of SF cannot be processed optimally, a nutritive transport medium such as the "Transgrow" bottle system or the JEMBEC plate should be inoculated at bedside and incubated for 18 hours to 24 hours before delivery or shipment to the laboratory. Failure to preincubate such nutritive transport media may affect adversely the subsequent recovery of gonococci by the laboratory.[14] Although inoculation of a special agar-overlap medium may allow a higher recovery of gonococci from SF in cases of disseminated gonococcal infection,[15] it is important to submit appropriate specimens from other body sites (skin lesions, anorectum, genitals, oropharynx) for culture and to inform the laboratory of all such specimens submitted.[14] Blood cultures should always be submitted in cases of suspected gonococcal arthritis since only 25% to 50% of SF cultures may be positive, whereas "gonococcal bacteremia is demonstrable in most patients . . . soon after the onset of gonococcal arthritis."[18]

Septic arthritis caused by *H. influenzae* type b, *N. meningitidis, Moraxella osloensis,* and *Streptobacillus moniliformis,* can mimic gonococcal arthritis. The Gram stain of the SF is rarely helpful in these cases, although a green-tinged, purulent SF is fairly characteristic of *H. influenzae* arthritis. Similar to *Neisseria, H. influenzae* is a fragile organism, and rapid handling of clinical specimens with avoidance of refrigeration is essential to ensure its survival; fortunately, direct bedside inoculation of the chocolate agar used for *Neisseria* with subsequent incubation at 35°C to 37°C in a moist CO_2 atmosphere should ensure survival of *H. influenzae* until the clinical specimens reach the laboratory.[17]

In addition to the organisms noted above, bacterial arthritis can be caused by various aerobic organisms, and at least 1 ml of SF should be submitted in a tightly sealed (screw-capped) sterile container (or syringe) for standard aerobic bacterial cultures. Such cultures are usually positive if previous antibiotic therapy has not been given, but, if antibiotics have been given, a history of treatment must be noted on the requisition form to guide the laboratory in processing the specimen. Any previous arthrocenteses or joint surgery should also be noted because these may have resulted in bacterial contamination of the joint space. In addition, the patient's medical history may provide clues as to the type of organism present (*e.g., Salmonella* species in sickle-cell disease). Blood cultures routinely should be submitted simultaneously with the samples of SF, although such cultures may be positive in only 50% or fewer of cases of acute suppurative arthritis. Any other obvious site of infection should also be cultured because it may yield the same pathogenic organism as the SF or, if the SF cultures are negative, may provide the only clue as to the etiologic agent of the arthritis. Because only a small number of organisms may be present and these may be trapped by clots, sterile sodium heparin (25–50 units/ml of SF) anticoagulant not containing preservatives may be added to samples of SF submitted for bacteriologic study.

If arthritis secondary to brucellosis is suspected (*e.g.,* a patient handling animal products), the laboratory should be notified before the arthrocentesis so that the sample of SF may be handled quickly and safely; multiple blood cultures should also be obtained, an acute phase serum should be collected for serologic study, and consideration should be given to obtaining a synovial biopsy at the time of arthrocentesis.

Although the gross appearance of the SF should always be noted, it is often a poor guide to the presence of septic arthritis. Purulent-appearing SF may be found in joints affected by crystal-induced synovitis or rheumatoid arthritis, and serous SF may be observed both in the early stages of severe suppurative arthritis and in partially treated bacterial arthritis.[8]

MYCOBACTERIA

Although tuberculous arthritis is no longer common in the United States, a separate sample of at least 5 ml of SF should be submitted for mycobacteriologic study, particularly if tuberculosis is suspected clinically. Special care must be taken with this specimen to avoid overgrowth by endogenous or exogenous bacteria, and it should be kept at 4°C if there is any delay in its processing. All personnel handling the potentially infected SF and the arthrocentesis equipment should take appropriate precautions, including the wearing of a surgical gown, mask, and gloves and the prevention of aerosol formation, which can spread the infected SF.

The preparation of an acid-fast stain of the sediment from centrifuged SF, preferably using one of the faster and more sensitive fluorochrome stains such as auramine–rhodamine, is the only part of the workup of suspected tuberculous arthritis that can be done with reasonable safety outside a properly equipped mycobacteriology laboratory.[19] Because of the small number of mycobacteria that may be present in the SF, such stains may not reveal any mycobacteria, but any observed acid-fast organism is of pathologic significance, and great care should be taken in examining such acid-fast stains. Even if acid-fast organisms are seen, culture is needed because tuberculous arthritis may be caused by mycobacteria other than *M. tuberculosis*.

ANAEROBIC BACTERIA

Although infection of joint spaces by anaerobic organisms is a relatively minor cause of septic arthritis (3–8% anticipated yield of anaerobes),[20] anaerobic cultures of SF should be performed if there is any laboratory or clinical suspicion of such infection. For anaerobic cultures, the SF may be submitted to the laboratory in a syringe and needle (stuck into a sterile rubber stopper) from which all air has been expelled. The anticipated transport time to the laboratory must be less than 30 minutes. This sample can also be used for general bacteriologic purposes. If longer transport periods are foreseen, the specimen should be submitted in a "gassed-out" collection tube that contains a prereduced, anaerobically sterilized transport medium. Alternatively, the SF may be inoculated into a commercially available sterile tube containing oxygen-free CO_2 or an anaerobic transport tube. Although the inoculation of culture media at the bedside has been recommended, it generally is impractical and probably unnecessary provided that the sample is collected properly, stored anaerobically, and transported expeditiously to the laboratory.[20, 21]

FUNGI

Fungal infection of a joint space can occur in various disseminated systemic mycoses either as a result of direct hematogenous spread to the joint space or by local extension from an adjoining cutaneous, subcutaneous, or bone lesion.[16]

Because of the slow growth of most systemic fungal pathogens, direct microscopic examination of appropriately pretreated (*e.g.*, potassium hydroxide) unstained wet mounts or stained smears of fresh SF (uncentrifuged or sediment) may allow a rapid diagnosis of the etiologic agent of disease. Of particular value for detection of the encapsulated yeast cells of *Cryptococcus neoformans* is a negative capsule stain, prepared by the addition of India ink or nigrosin to a smear of sediment from centrifuged (1000 × *g*, 15 minutes) SF.[22]

Synovial fluid suspected of containing fungi should be inoculated, if possible, into a culture medium at the bedside within 1 hour of collection. For this purpose, Sabouraud dextrose agar slants in screw-capped tubes or bottles are preferred. Sabouraud agar that contains the antibiotics chloramphenicol and cycloheximide (Mycosel agar) is useful in preventing bacterial overgrowth if prolonged storage or transport of the samples (*e.g.*, to an outside mycology laboratory) is anticipated. Such antibiotics are inhibitory, however, to the growth of a number of fungi, and agar that contains such antibiotics should not be used exclusively if infection owing to one of the susceptible fungi is suspected; if they are used, another sample of uninoculated SF should be retained.[23]

Generally, if delays in processing uninoculated SF are unavoidable or if the SF is to be kept for reference purposes after primary inoculation, it should be stored at 30°C, not at 4°C. Although refrigeration is not intrinsically harmful to any fungus that may be present, it will delay the rate of growth and prolong the time needed to make a specific identification. Bacterial overgrowth in SF by either endogenous or exogenous organisms, which will be minimized at 4°C but accelerated at 30°C, may be prevented by the addition of a combination of broad-spectrum antibiotics such as penicillin (20 units/ml of SF) and streptomycin (40 units/ml of SF); antibiotics must not be added to SF if infection by the "fungal-like" bacteria *Actinomyces* species or *Nocardia* species is suspected.

VIRUSES

Although viral arthritis is a relatively uncommon cause of joint disease, it has been reported in association with viral diseases including acute hepatitis B, rubella, mumps, and infection by the herpes family of viruses.[24–27] In cases of suspected viral arthritis, the laboratory virologist should be consulted before the arthrocentesis, and all appropriate precautions should be taken during collection and handling of the SF and disposal of the used arthrocentesis supplies.

Although there are no firm guidelines as to the handling of SF for virology, by analogy with the CSF and other normally sterile body fluids, at least 1 ml to 3 ml of SF should be placed in a dry, sterile, screw-capped, glass container that does not have a virus transport medium. Many pathogenic viruses are quite labile at ambient temperature, and if the appropriate tissue culture system cannot be inoculated at the bedside, then the sample must be transported to the laboratory as expeditiously as possible. In particular, because freezing can destroy the infectivity of some pathogenic viruses (*e.g.*, cytomegalovirus, herpes simplex virus, mumps virus, rubella virus), the specimen of SF should be transported at ambient temperature (about 23–25°C) or, if a delay of not more than a few hours is anticipated, kept at 4°C (melting ice). Samples of SF that require long transit times (shipment to an outside laboratory) ideally should be frozen quickly ("snap") at −70°C or lower (Dry Ice, Dry Ice-alcohol, liquid nitrogen) and shipped at this temperature; however, fluctuations in temperature with concomitant freezing and thawing must be avoided. Because shipment at such low temperature is not often feasible, such specimens probably should be shipped unfrozen at 4°C in a tightly sealed, screw-capped vial (with adhesive tape covering the cap) in an appropriate insulated container. If delays in transit occur, the samples should be kept refrigerated at 4°C and should not be placed in a deep freezer. Detailed instructions for the shipment of specimens that contain viruses are given in a World Health Organization monograph.[28]

The shedding of viruses into body fluids often decreases rapidly after onset of illness, and specimens of SF thus should be collected as soon as possible after clinical onset of disease (acute-phase specimen). In addition, although the etiologic virus may be isolated from the SF,[25, 27] it may also be isolated from other clinical specimens such as throat swabs, vesicle fluid, urine, and stool, and such material must be collected properly and submitted with the specimen of SF.[29] In addition, because of the usefulness of serologic tests to diagnose viral disease, 5 ml to 10 ml of serum should be collected during the acute phase of the viral illness and kept frozen (−20– −70°C) until needed.

Finally, because viruses require living cells for replication, viral titers decline rapidly after death, and postmortem samples of SF for virologic study must be collected aseptically as quickly as possible after death. Because bacterial overgrowth, which will interfere with viral isolation, may occur rapidly after death, it may be advisable to add antibiotics to such specimens.

MICROSCOPIC EXAMINATION

After the samples for microbiologic analysis and bedside analysis have been removed, a specimen of SF for microscopic analysis is prepared by transferring at least 5 ml of freshly aspirated SF into a tube that contains 25 to 50 units of sodium heparin per milliliter of SF as an anticoagulant and rapidly mixing. Although 10% EDTA in normal saline has also been used as an anticoagulant, the use of crystalline anticoagulants such as EDTA salts and oxalate should be avoided because their use may lead to an erroneous diagnosis of crystal-deposition disease.[8, 30, 31] A clot should never form in the anticoagulated tube regardless of the nature of the arthropathy; if observed, it is indicative either of untoward delay in transferring the aspirated SF to the anticoagulant or of failure to mix the SF and the anticoagulant thoroughly.

As noted, the results of a bedside "wet drop" preparation must be communicated to the laboratory. In addition, because prolonged microscopic examination of SF may be necessary to find even a few monosodium urate (MSU) crystals in case of acute gouty arthritis,[32] the clinical suspicion of a crystal-induced arthritis must be put on the sample requisition form so that laboratory personnel will undertake a prolonged study of the SF. Other related information that must be noted on the requisition form is a history of any previous intraarticular injection of corticosteroid ester suspensions (which may persist in the SF for 1 month or longer). Such crystalline material may be confused with MSU or calcium pyrophosphate dihydrate (pseudogout) and may itself result in a "crystal-induced" synovitis that lasts 1 to 3 days.[33] Any previous open surgery on the joint should be noted because of possible contamination of the joint space by talc or metal particles.[34]

Ideally, samples of SF for cytologic study should be examined unfixed and as soon as possible after collection to minimize changes in cellular structure.

Similar to other body fluids that are relatively high in protein, cells in SF should be relatively well preserved for up to 48 hours to 72 hours if the SF is refrigerated at 4°C,[8, 35] although it has been reported that white cells, lupus erythematosus cells, rheumatoid arthritis cells, crystals, and cartilage fragments in SF retain their structure for more than 1 year by "slow freezing in the ordinary refrigerator freezing department (0 to −10°C)."[36] However, if specimens of SF cannot be examined within 48 hours after collection, they should be prefixed within 1 hour of collection by addition of an equal volume of 50% ethanol to the fluid. This is particularly important if a diagnosis of metastatic or primary cancer of the joint is based on the exfoliative cytology of the SF.[35, 37] The cellular morphology of prefixed SF will remain unchanged for at least 1 week.

SEROLOGY

An additional 5 ml to 10 ml of SF should be placed in a tube not containing an anticoagulant and allowed to stand at ambient temperature for 10 to 15 minutes. The tube should then be centrifuged to remove any clot that may have formed or any cellular material present, and the supernatant used for serologic (immunologic) studies. In particular, if complement determinations are to be done, the supernatant should either be analyzed within 2 hours of collection or frozen immediately and stored at −70°C until needed; concurrently obtained samples of serum (for reference purposes) must also be kept at −70°C until analyzed.[38] Finally, although many complement assays can be done on serum and the supernatant from coagulated SF, some assays require plasma and anticoagulated (EDTA) SF,[38] and hence an awareness of the specific complement assay to be done is needed before the SF (and blood) are processed and submitted to the laboratory.

REFERENCES

1. McCarty DJ: Synovial fluid. In McCarty DJ (ed): Arthritis and Allied Conditions: A Textbook of Rheumatology, 9th ed, pp 51–69. Philadelphia, Lea & Febiger, 1979
2. Ropes MW, Bauer W: Synovial Fluid Changes in Joint Disease. Cambridge, Harvard University Press, 1953
3. Cohen AS, Brandt KD, Krey PR: Synovial fluid. In Cohen AS (ed): Laboratory Diagnostic Procedures in the Rheumatic Diseases, 2nd ed, pp 1–62. Boston, Little, Brown & Co, 1975
4. Rippey JH: Synovial fluid analysis. Lab Med 10:140–145, 1979
5. Cracchiolo A III: Joint fluid analysis. Am Family Physician 4:87–94, 1977
6. Miller JA Jr: Joint paracentesis from an anatomic point of view. I. Shoulder, elbow, wrist, and hand. Surgery 40:993–1006, 1956
7. Miller JA Jr: Joint paracentesis from an anatomic point of view. II. Hip, knee, ankle, and foot. Surgery 41:999–1011, 1957
8. Currey HLF, Vernon-Roberts B: Examination of synovial fluid. Clin Rheum Dis 2:149–177, 1976
9. Goldenberg DL, Brandt KD, Cohen AD: Rapid, simple detection of trace amounts of synovial fluid. Arthritis Rheum 16:487–490, 1973
10. Krauss DS, Aronson MD, Gump DW, Newcombe DS: *Hemophilus influenzae* septic arthritis: A mimicker of gonococcal arthritis. Arthritis Rheum 17:267–271, 1974

11. Hasselbacher P: Measuring synovial fluid viscosity with a white blood cell diluting pipette. Arthritis Rheum 19:1358–1362, 1976
12. Halverson P, McCarty DJ: Unpublished data cited in Reference 1, p 56
13. Bluhm GB, Riddle JM, Barnhardt MI, Duncantt, Sigler JW: Crystal dynamics in gout and pseudogout. Med Times 97(6):135–143, 1969
14. Morello JA, Bohnoff M: *Neisseria* and *Branhamella*. *See* Reference 19, pp 111–130
15. Holmes KK, Gutman LT, Belding ME, Turck M: Recovery of *Neisseria gonorrhoeae* from "sterile" synovial fluid in gonococcal arthritis. N Engl J Med 284:318–320, 1971
16. Ansell BM: Infective arthritis. In Scott JT (ed): Copeman's Textbook of the Rheumatic Diseases, 5th ed, pp 808–829. Edinburgh, Churchill Livingstone, 1978
17. Kilian M: Haemophilus. *See* Reference 19, pp 330–336
18. Holmes KK, Counts GW, Beaty HN: Disseminated gonococcal infection. Ann Intern Med 74:979–993, 1971
19. Runyon EH, Karlson AG, Kubica GP, Wayne LG (revised by Sommers HM, McClatchy JK): Mycobacterium. In Lennette EH, Ballows A, Hausler WJ Jr, Truant JP (eds): Manual of Clinical Microbiology, 3rd ed, pp 150–179. Washington DC, American Society for Microbiology, 1980
20. Talley FP, Bartlett JG, Gorback SL: Practical guide to anaerobic bacteriology. Lab Med 9(9):26–35, 1978
21. Allen SD, Siders JA: Procedures for isolation and characterization of anaerobic bacteria. *See* Reference 19, pp 397–417
22. Rinaldi MG: Cryptococcosis. Lab Med 13:11–19, 1982
23. Larsh HW, Goodman NL: Fungi of systemic mycoses. *See* Reference 19, pp 577–594
24. Malawista SE, Steere AC: Viral arthritis. In Kelley WN, Harris ED Jr, Ruddy S, Sledge CB (eds): Textbook of Rheumatology, pp 1586–1601. Philadelphia, WB Saunders, 1981
25. Priest JR, Urick JJ, Groth KE, Balfour HH Jr: Varicella arthritis documented by isolation of virus from joint fluid. J Pediatr 93:990–992, 1978
26. Duffy J: Viral arthritis: State of the art 1981. Mayo Clin Proc 56:579, 1981
27. Friedman HM, Pincus T, Gibilisco P, Baker D, Glazer JP, Plotkin SA, Schumacher HR: Acute monoarticular arthritis caused by herpes simplex virus and cytomegalovirus. Am J Med 69:241–247, 1980
28. Madeley CR: Guide to the Collection and Transport of Virological Specimens. Geneva, World Health Organization, 1972
29. Ray CG, Hicks MJ: Laboratory diagnosis of viruses, rickettsia, and chlamydia. In Henry JB (ed): Todd·Sanford·Davidsohn Clinical Diagnosis and Management by Laboratory Methods, 16th ed, pp 1814–1879. Philadelphia, WB Saunders, 1979
30. Phelps P, Steele AD, McCarty DJ: Compensated polarized light microscopy. JAMA 203:508–512, 1968
31. Schumacher JR: Intracellular crystals in synovial fluid anticoagulated with oxalate. N Engl J Med 274:1372–1373, 1966
32. Schumacher HR, Jiminez SA, Bison T, Paseual E, Tragcoff R, Dorwart BB, Reginato AJ: Acute gouty arthritis without urate crystals identified on initial examination of synovial fluid. Arthritis Rheum 18:608–612, 1975
33. Kahn CB, Hollander JL, Schumacher HR: Corticosteroid crystals in synovial fluid. JAMA 211:807–809, 1970
34. Kitrou R, Schumacher HR Jr, Sbarbaro JL, Hollander JL: Recurrent hemarthosis after prosthetic knee arthroplasty: identification of metal particles in the synovial fluid. Arthritis Rheum 12:520–527, 1969
35. Bales CE, Durfee GR: Cytological techniques. In Koss LG: Diagnostic Cytology and Its Histopathological Bases, 3rd ed, pp 1187–1266. Philadelphia, JB Lippincott, 1979
36. Backer GB, Rodriquez CE, Koehl CW: Microscopic evaluation of frozen synovial fluid. Arthritis Rheum 8:429, 1965
37. Naib ZM: Cytology of synovial fluids. Acta Cytol 17:299–309, 1973
38. Hunder GG, McDuffie FC, Mullen BJ: Activation of complement components C3 and factor B in synovial fluids. J Lab Clin Med 89:160–171, 1977

Cerebrospinal Fluid

Cerebrospinal fluid (CSF), found in the ventricles of the brain and the subarachnoid space, furnishes physical support for the brain, helps to maintain chemical homeostasis in the central nervous system (CNS), supplies an excretory function for the brain, and provides a channel for transportation within the CNS. The diagnosis and proper treatment of CNS disease may depend on the results of laboratory examination of the CSF; thus this fluid must be obtained and handled properly both before and during laboratory analysis.

COLLECTION AND PROCESSING

COLLECTION SITES

The CSF is normally sterile and can be obtained by lumbar puncture (LP) or, less commonly, by cisternal, cervical, or ventricular puncture; each of these procedures must be performed aseptically by a physician experienced in the procedure and aware of its indications and potential complications.[1–3] If CSF is to be removed at the time of a diagnostic procedure (*e.g.*, myelography or pneumoencephalography) or intrathecal therapy, it generally should be obtained before the injection of diagnostic or therapeutic agents into the subarachnoid space to avoid any confusion that may result from inflammatory reaction caused by the injected substance.

REQUISITION FORM

The requisition accompanying the sample should have written on it any known information (even if presumptive) about the nature of the patient's disease, details of any previous therapy (surgery, antibiotics, chemotherapy, radiotherapy), and results of other recent LPs.

Part of the proper evaluation of the CSF includes prompt (within 1 minute) and accurate measurement (not estimation) of the initial or opening pressure (before removing any fluid) in a relaxed and properly positioned patient. This bedside measurement should be recorded on the patient's chart and on the laboratory request form.

COLLECTION TUBES AND LABELS

Normally from an adult three 3- to 4-ml samples of CSF should be collected in sterile, preferably screw-capped plastic tubes that are sequentially labeled #1 (for chemistry and serology), #2 (for microbiology), and #3 (for microscopic analysis that includes cell count and cytology). Even if not requested, it is good to collect one or two extra tubes of CSF to avoid the need for an additional LP if additional studies are needed. Up to 20 ml of CSF (about 15% of the estimated total CSF volume) may be removed safely from an adult in the absence of an increased opening pressure or a marked fall in the CSF pressure on removal of the fluid. Proportionally smaller volumes must be removed in newborns, infants, and children (whose total CSF volumes are considerably less than those of adults), although at least 1 ml to 2 ml should always be collected. Even though the risk of development of a post-LP headache may be related to the amount of CSF removed, this risk must be weighed against the diagnostic advantages that may be accrued by removal of relatively large amounts of spinal fluid.[2]

If only a small amount of CSF has been collected (*e.g.*, from a neonate) in a case of suspected meningitis, it should first be sent, undivided, to the microbiology laboratory so that it can be handled aseptically and cultures can be begun immediately. Samples of CSF for cytology, serology, and chemical analysis can be removed at this time, or, if the sample of CSF is very small, it can be centrifuged, the sediment used for microbiology, and the supernatant used for the additional studies. As already noted, however, relatively large (at least 5 ml and preferably 10 ml) samples of CSF are needed in most instances for the identification of fungi and mycobacteria, and 20 ml (or much more) of CSF may be needed for a cytologic identification of tumor cells.

CSF APPEARANCE

Any change from the normal appearance of the CSF (colorless, crystal-clear, and of a viscosity similar to serum) may indicate CNS abnormalities. Increases in viscosity are rare but can occur with metastatic, mucin-secreting adenocarcinomas. An increased number of cells (pleocytosis) in the CSF will produce an observable turbidity or cloudiness at a level of about 400 cells/μl, and such turbidity can be graded semiquantitatively from 0 (crystal clear) to 4+ (newsprint not readable through tube). These estimates must be made using 1 ml to 2 ml of CSF in a glass tube. Plastic collection tubes should not be used for this measurement.

Although not producing turbidity, CSF pleocytosis of less than about 400 cells/μl can be detected at bedside by the Tyndall effect. In this test, the top of a clean glass test tube that contains CSF is held in direct sunlight against a dark background with one hand while the bottom of the tube is flicked briskly with a finger of the other hand; pleocytosis results in "snowy" or "sparkling" appearance in the CSF because of incident light scatter. With this bedside test, white or red blood cell counts as low as 4 cells/μl can be identified with high specificity (absence of false-positives).[4]

The color of CSF, both during collection, and after centrifugation, must be examined carefully because the differential diagnosis of a subarachnoid hemorrhage (SAH) depends, in part, on such observations.

Nonhomogeneous mixing of blood and CSF in the manometer, used to measure the opening and closing pressure during the LP, followed by visible clearing with time or a decrease in blood between the first and third tubes (confirmed by serial red cell counts or microhematocrits on the collected tubes of CSF) is evidence of a traumatic tap. However, a decreasing hematocrit from tubes 1 to 3 may be observed in the presence of an SAH because of postural layering of cells in the CSF.

Further bedside differentiation between an SAH (or intracerebral hemorrhage) and a traumatic tap can be accomplished by the immediate (within 1 hour of collection) centrifugation of the CSF and the observation of xanthochromia in the supernatant. Xanthochromia, which literally means yellow color but which clinically may be used to refer to a spectrum of colors from yellow to orange to pink, is best evaluated by comparing at least 1 ml to 2 ml of CSF supernatant to an equal volume of water in an identical tube (viewed down the tubes' long axes). Its presence indicates that the blood has been in the CSF for at least several hours and suggests that such blood arose from a preexisting hemorrhage. Xanthochromic CSF may be obtained in the absence of intrinsic bleeding if the CSF protein is greater than 150 mg/dl; if the patient is jaundiced (hyperbilirubinemic), hypercarotenemic, or has a meningeal melanoma; or if the LP has been performed 2 to 5 days after a traumatic tap. The CSF in normal neonates is usually xanthochromic owing to a combination of hyperbilirubinemia and increased CSF protein.[5]

A clot or surface pellicle (film) forming in a tube of CSF indicates an abnormality. It may form in the CSF from a very bloody traumatic tap in which there are more than about 200,000 red cells/μl. The blood from an SAH will be defibrinogenated and will not clot unless the hemorrhage has recently occurred or is continuing at the time of the LP.

MICROBIOLOGIC STUDIES

BACTERIA

Although rapid processing of CSF specimens is always indicated, immediate processing of CSF specimens is necessary in cases of suspected CNS infections. Bacterial meningitis can be rapidly fatal or result in serious, permanent neurologic deficits if untreated or inadequately treated, and prompt identification of the causative agent is therefore needed. Processing of CSF specimens for microbiology studies should start at the bedside and include the preparation of a Gram-stained smear of centrifuged ($1000 \times g$ for 15 minutes) sediment on an alcohol-washed and flamed sterile glass slide and the inoculation of appropriate culture media for fastidious pathogenic organisms such as *N. meningitidis* (Thayer–Martin chocolate agar or New York City medium plates, Transgrow bottles if extended—2 to 3 days—transport is anticipated) and *H. influenzae* (Thayer–Martin chocolate agar). Such inoculated samples of CSF should be maintained at 35°C to 37°C in a humid CO_2 (3–10%) atmosphere (*e.g.*, in a candle jar) during transportation to the laboratory to maximize the recovery of such fragile organisms that may not survive prolonged storage or temperature variation. It is particularly important to protect these specimens from drying and temperature extremes (especially cold) while in transit to the laboratory.

In addition to any cultures inoculated at bedside, at least 3 ml to 4 ml of CSF (tube #2) should be submitted in a sterile, screw-capped tube for routine aerobic cultures. A small amount of sterile sodium heparin (25–50 units/ml of SF) not containing preservatives may be added to the CSF to prevent the trapping of bacteria by any clots that may form.

Blood cultures, which may be positive in 40% to 60% of cases of meningococcal, pneumococcal, or *H. influenzae* meningitis, should always be obtained in cases of suspected bacterial meningitis because they may provide the only means of identifying the agent of infection if the CSF cultures are negative. If a CSF lactic acid assay is used to aid in the early diagnosis of bacterial meningitis, then the specimen must be brought to the laboratory immediately after the LP. "Significant changes in lactic acid concentration in cerebrospinal specimens obtained from a patient with bacterial meningitis can occur within 15 minutes of incubation at room temperature."[6] If such measurements cannot be done promptly, however, the original lactic acid concentration can be preserved by freezing the specimen immediately and storing it at −20°C.

In mycobacterial infection of the CNS, relatively few organisms may be shed into the CSF, and relatively larger amounts (5 ml or more) of CSF may be needed. If possible, the specimen should be obtained in a separate tube and taken to the laboratory as quickly as possible. If there is to be any delay in its transport to the laboratory, the specimen should be refrigerated at 4°C. A small amount of sterile sodium heparin may be added to the CSF specimen to prevent the trapping of any mycobacteria in clots.

If delays are anticipated during shipment of the specimen to the laboratory, the CSF may be inoculated into either Middlebrook 7H9 broth or another noninhibitory culture medium to enhance mycobacterial growth and to minimize the time required to make a positive identification of the organism. Because of possible inhibitory effects on mycobacterial growth, selective culture media that contain antibiotics should not be used. Overgrowth by coexisting bacteria or fungi usually is not a problem with a normally sterile body fluid such as CSF unless a polymicrobial infection is present.

Generally, all specimens of CSF suspected of containing mycobacteria should be processed only by properly trained technologists in a properly equipped laboratory.[7] The only procedure that should be performed outside such a laboratory is the preparation of an acid-fast stain of centrifuged CSF sediment.[7] Because of the small number of mycobacteria that may be present, such stains may not reveal any mycobacteria in more than half of all cases of mycobacterial tuberculosis. Any acid-fast organism is, however, of pathologic significance.

If a surface pellicle is present on the CSF, a small piece should be removed aseptically for an acid-fast stain and the remainder inoculated into various culture media. Although mycobacterial meningoencephalitis usually is diagnosed by culture, microscopy may be positive when cultures are negative.

Although infection of the meninges by anaerobic bacteria is an infrequent cause of meningitis (less than 1% anticipated yield of anaerobes in CSF), such organisms are a major cause of brain abscesses (89% incidence) and subdural empyema (about 50% incidence).[8, 9] Anaerobic cultures of CSF should be performed, especially if there is any laboratory or clinical suspicion of such anaerobic infection (*e.g.*, spore-forming rods seen in Gram-stained CSF, foul-smelling CSF, CT scan compatible with brain abscess, CNS abnormalities with coexistent anaerobic infection elsewhere in the body).

The diagnosis of such an anaerobic infection depends on the proper handling of the specimen which may include a brain aspirate as well as CSF. If the anticipated transport time to the laboratory is less than about 30 minutes, the CSF may be submitted for anaerobic cultures in a syringe and needle (stuck into a sterile cork or rubber stopper) from which all air has been expelled. (Such specimens are also suitable for general bacteriologic cultures.) If longer transit times are foreseen, the sample should be submitted in a "gassed-out" collection tube that contains a prereduced, anaerobically sterilized transport medium. Alternatively, the CSF may be inoculated into either a commercially available sterile tube containing oxygen-free CO_2 or an anaerobic transport tube.

Because of the potentially grave complications of bacterial meningitis, avoidance of both false-negative and false-positive identification of the etiologic agent is important. False-negative findings may lead to erroneous diagnosis of aseptic meningitis and result in withholding of appropriate antibiotic therapy, and false-positive findings may result in use of inappropriate antibiotic therapy and failure to continue to search for the correct cause of the CNS disease.

False-negative results may arise from inappropriate collection or mishandling of the clinical sample and from previous inadequate antibiotic therapy that has suppressed bacterial growth. For this reason, any previous drug therapy should be indicated on the sample requisition form to guide the laboratory evaluation of the CSF.[10]

False-positive Gram stains of centrifuged CSF may result from nonviable bacteria in the collection tubes, on glass slides not precleaned with alcohol, and in unfiltered Gram-stain solutions.[3, 11]

FUNGI

Specimens of CSF to be examined for fungi generally can be collected in the same way as those for bacteriologic examination, and ideally should be obtained in separate tubes. Because only small numbers of fungi are usually seeded from infected tissue into the CSF, a relatively large amount (at least 5–10 ml) of CSF should be collected, especially in cases of suspected infections by *Cryptococcus neoformans*. In cases of possible cryptococcal meningitis, direct bedside examination of a smear of centrifuged (1000 $\times$ *g*, 15 minutes) CSF sediment to which India ink or, preferably, nigrosin has been added as a negative-capsule stain may allow rapid detection of the encapsulated *C. neoformans* in about 50% of cases.[12] Alternatively, this yeast may be detected rapidly by microscopic examination of alcian blue or mucicarmine-stained dry smears of the centrifuged CSF sediment.[13]

In cases of suspected CNS fungal disease, samples of CSF should not be stored before inoculation of the primary culture media. Ideally, inoculation should be done at the bedside within 1 hour of collection. For this purpose, Sabouraud dextrose agar slants in screw-capped tubes or bottles are preferred, although Sabouraud agar that contains chloramphenicol and cycloheximide (Mycosel agar) is useful in preventing bacterial overgrowth if prolonged storage or transport is anticipated. Cycloheximide also inhibits the growth of *C. neoformans,* and media containing this antibiotic should not be used exclusively for the isolation of fungi from CSF. If such media are used, another sample of uninoculated CSF should be retained and stored at 30°C.[14]

If delays in processing of uninoculated CSF are unavoidable, or if the CSF is to be kept for reference purposes after inoculation of the primary culture medium, it should be stored at 30°C, not at 4°C. Although such refrigeration will not be harmful to any fungus present, it will impede its rate of growth and unnecessarily prolong the time required for specific identification. Bacterial overgrowth in such CSF may be prevented by the addition of various antibiotic combinations such as 20 units/ml of penicillin and 40 units/ml of streptomycin (final CSF concentrations).[14] Antibiotics must not be added to specimens of CSF to be examined for the "fungal-like" bacteria *Nocardia* species or *Actinomyces* species.

Microscopic evaluation of the number and morphology of the fungi at the time of collection can be performed after fixing part of the CSF either as a dry smear (*cf.*, detection of *C. neoformans*) on a microscope slide or by the addition of 10% formalin.[13]

AMOEBA

With suspected primary amoebic meningoencephalitis caused by a pathogenic member of the genera *Naegleria* or *Acanthamoeba*, a parasitologist should be contacted and a sample of CSF specifically for parasitology obtained aseptically and placed in a tightly sealed, screw-capped container. The specimen must be kept at ambient (24–28°C) temperature and under no circumstances should be refrigerated or frozen because this will decrease the number of viable parasites in the specimen. The sample must be brought expeditiously to a laboratory equipped with a biologic safety hood for isolation, identification, and culture of the organism.[15] All personnel handling such CSF (and contaminated supplies) should wear surgical masks and gloves.

VIRUSES

With suspected viral infection of the CNS, the laboratory virologist should be consulted before samples of the CSF are collected, and all appropriate precautions should be taken during the collection and handling of the CSF and disposal of the used LP equipment, especially if there is possible CNS infection by a "slow" virus, such as that which may be involved in Creutzfeldt–Jakob disease.[16]

About 1 ml to 3 ml of CSF, should be placed in a dry, sterile, screw-capped glass container that does not have a virus transport medium, is needed for virologic study. Because many pathogenic viruses are extremely labile in CSF at ambient temperature, rapid transport of the specimen to the laboratory with immediate inoculation (at the bedside, if possible) of the appropriate tissue culture systems is essential.

The shedding of viruses into biologic fluids such as the CSF often decreases rapidly after the onset of illness. Samples of CSF should therefore be collected as soon as possible after the appearance of signs and symptoms of disease (acute-phase specimen). Although some viruses may be isolated from the CSF in viral meningoencephalitis, they are equally or more likely to be obtained from other clinical specimens such as feces, blood, throat swabs, urine, vesicle fluid, and brain tissue. Such material must be collected properly and submitted with the CSF.[17] Because of the usefulness of serologic tests in the diagnosis of viral disease, 5 ml to 10 ml of serum should be collected during the acute phase of the viral illness and kept frozen (−20– −70°C) until needed.

Because viruses need living cells for growth, viral titers decline rapidly after death. Postmortem samples of CSF (and brain tissue) for virologic study must be collected as quickly and as aseptically as possible because bacterial overgrowth may occur rapidly after death and will interfere with viral isolation. This problem can be minimized by the addition of antibiotics (penicillin, 500 units/ml; genatmicin, 50 μg/ml; and amphotericin B, 10 μg/ml) to the CSF.

Freezing can destroy the infectivity of some viruses (*e.g.*, arboviruses, herpes simplex virus, mumps virus). The CSF specimen should be transported at ambient temperature or, if a delay of more than a few minutes (but less than several hours) is anticipated, kept at 4°C (melting ice). Samples of CSF that require long transit times usually should be frozen quickly ("snap") at −70°C or lower (Dry Ice, Dry Ice-alcohol, liquid nitrogen) and shipped at this temperature. Fluctuations in temperature, with concomitant freezing and thawing, must be avoided. Because shipment at such low temperatures often is not feasible, it is probably best to ship specimens for viral isolation *unfrozen* at 4°C in a tightly sealed, screw-capped vial (with adhesive tape covering the cap) in an appropriate insulated container; if delays in transit occur, the specimens should be kept refrigerated at 4°C and should not be stored in a deep freezer. Detailed instructions of the proper procedures for shipping specimens of CSF that contain viruses are stated in a World Health Organization monograph.[18]

CHEMICAL TESTS

For proper evaluation of the CSF glucose and protein levels, a simultaneously drawn sample of venous blood (serum) should be obtained and sent to the laboratory. Because the glucose level may be used in making a diagnostic decision, if possible the patient should have been fasting for at least 4 hours before the LP to ensure full equilibration between the blood and CSF glucose. Any intravenous infusions of glucose should be stopped or replaced by saline at least 1 to 2 hours before an LP.[21] CSF specimens for glucose measurement must be promptly analyzed because cells in the fluid may metabolize the fluid's glucose, resulting in a falsely low glucose level.

Although many special tests (enzymes, hormones, biogenic amines) can be done on CSF, they are often of limited clinical value.[1] If these are ordered, a blood (serum) sample must be drawn simultaneously for determination of the concurrent blood value of the substance in question.

If CSF is to be tested for myelin basic protein (MBP) for the evaluation of myelinolysis (*e.g.*, in multiple sclerosis), 3 ml of CSF is needed, and this may be stored at −20°C until used. (Myelin basic protein itself appears to be stable at 25°C in CSF for at least 1 week).

If CSF *p*H, P_{CO_2}, and bicarbonate are to be measured, the sample should be obtained anaerobically in a heparinized glass syringe, maintained at 0°C (ice bath) under anaerobic conditions (tip of needle or syringe sealed), and analyzed within 20 minutes of collection if possible. An arterial blood gas sample should be obtained simultaneously for reference purposes.

CYTOLOGY

Ideally, CSF cytologic studies should be done on unfixed samples to minimize morphologic changes, especially if membrane filters (*e.g.*, Millipore, Nuclepore) are used for cellular concentration. Because of the relatively low protein content of CSF, however, the cellular morphology of unfixed CSF will remain unchanged for only 1 to 2 hours after collection, even if refrigerated. Both red and white blood cells (especially polymorphonuclear granulocytes) will begin to undergo cytolysis after a few hours, resulting in spurious xanthochromia and a falsely low white cell count, respectively. In addition, morphologic alterations in exfoliated malignant cells in the CSF may make a diagnosis based on such cells difficult or impossible.

If CSF must be stored for any length of time, it should be prefixed within 1 hour of collection in either 50% ethanol (1:1 v/v), which is the best universal fixative,[19] or in 4% formalin. Cerebrospinal fluid fixed in this manner may be stored almost indefinitely at ambient temperature and may be transported without loss of morphologic detail. If nonfixed CSF must be transported, it should be kept in plastic-coated containers because mononuclear phagocytes have a high affinity for glass and changes occur in the white blood cell differential count of CSF stored in glass containers.[20]

REFERENCES

1. Fishman RA: Cerebrospinal Fluid in Diseases of the Nervous System, Philadelphia, WB Saunders, 1980
2. Tourtellotte WW, Haerer AF, Heller GL et al: Post-Lumbar Puncture Headaches. Springfield, Illinois, Charles C Thomas, 1964
3. Weinstein RA, Bauer FW, Hoffman RD, Tyler PG, Anderson RL, Stamm WE: Facitious meningitis. Diagnostic error due to nonviable bacteria in commercial lumbar puncture trays. JAMA 233:878–879, 1975
4. Simon RP, Abele JS: Spinal-fluid pleocytosis estimated by the Tyndall effect. Ann Intern Med 89:75–76, 1978
5. Naidoo BT: The cerebrospinal fluid in the healthy newborn infant. S Afr Med J 42:933–935, 1968
6. Brook I: Stability of lactic acid in cerebrospinal fluid specimens. Am J Clin Pathol 77:213–216, 1982
7. Runyon EH, Karlson AG, Kubica GP et al (revised by Sommers MH, McClatchy JK): Mycobacterium. In Lennette EH, Ballows A, Hausler WJ Jr et al (eds): Manual of Clinical Microbiology, 3rd ed, pp 150–179. Washington, DC, American Society for Microbiology, 1980
8. Allen SD, Siders JA: Procedures for isolation and characterization of anaerobic bacteria. *See* Reference 7, pp 397–417
9. Talley FP, Bartlett JG, Gorback SL: Practical guide to anaerobic bacteriology. Lab Med 9(9):26–35, 1978
10. Converse GM, Gwaltney JM Jr, Strassburg DA, Hendley JO: Alteration of cerebrospinal fluid findings by partial treatment of bacterial meningitis. J Pediatr 83:220–225, 1973
11. Muscher DM, Schell RF: False-positive Gram-stains of cerebrospinal fluid. Ann Intern Med 79:603–604, 1973
12. Rinaldi MG: Cryptococcosis. Lab Med 13:11–19, 1982
13. Silva–Hutner M, Cooper BH: Yeasts of medical importance. *See* Reference 7, pp 562–576
14. Larsh HW, Goodman NL: Fungi of systemic mycoses. *See* Reference 7, pp 557–594
15. Visvesvara GS: Free-living pathogenic amoebae. *See* Reference 7, pp 704–708
16. Traub RD, Gajdusek DC, Gibbs CJ Jr: Precautions in autopsies in Creutzfeldt–Jakob disease. Am J Clin Pathol 64:287, 1975
17. Ray CG, Hicks MJ: Laboratory diagnosis of viruses, rickettsia, and chlamydia. In Henry JB (ed): Todd•Sanford•Davidsohn Clinical Diagnosis and Management by Laboratory Methods, 16th ed, pp 1814–1879. Philadelphia, WB Saunders, 1979

18. Madeley CR: Guide to the Collection and Transport of Virological Specimens. Geneva, World Health Organization, 1977
19. Bales CE, Durfee GR: Cytological techniques. In Koss LG: Diagnostic Cytology and Its Histopathological Bases, 3rd ed, pp 1187–1266. Philadelphia, JB Lippincott, 1979
20. Dayan AD, Stokes MI: Rapid diagnosis of encephalitis by immunofluorescent examination of cerebrospinal-fluid cells. Lancet 1:177–179, 1973
21. Calabrese VP: The interpretation of routine CSF tests. Va Med Monthly 103:207–209, 1976

Amniotic Fluid

During the past decade information about amniotic fluid constituents and their concentrations has greatly increased. The clinical application of this information has become widespread owing to the development of safe techniques for obtaining amniotic fluid samples by amniocentesis. For amniocentesis, a needle attached to a syringe is inserted through the abdominal wall into an intrauterine pool of amniotic fluid, and the fluid is aspirated. If the fluid is bloody, the initial syringe is replaced, and additional fluid is aspirated to obtain a sample that contains considerably less blood or no blood. An amniocentesis can be performed from the 12th to the 42nd week of gestation, the exact time depending on the needed clinical information. Before an amniocentesis is performed, ultrasound and physical examination findings are used to determine the location of all fetal parts and fetal presentation.

The three general reasons for analyzing amniotic fluid are in utero detection of genetic disease of the fetus, detection of fetal jeopardy, and determination of fetal maturity.

Many inherited diseases can be detected in utero by biochemical analysis of amniotic fluid supernatant, or by cytochemical examination of uncultured or, more commonly, cultured amniotic fluid cells. Chromosomal evaluations can detect abnormalities such as Down's syndrome (trisomy 21). Fetal sex may be determined to identify fetuses at risk for sex-linked disease.

Certain maternal conditions may adversely affect the fetus. These include Rh isosensitization, preeclampsia, eclampsia, and diabetes mellitus. In these cases, evaluation of fetal status may be needed to determine if the fetus is in jeopardy. Amniotic fluid analysis can provide this information.

If the fetus is at risk or if delivery by cesarean section is necessary, an indicator of maturity, especially that of the lungs, is required. A number of tests are available for this purpose, and the most useful are the lung maturity tests (*i.e.*, lecithin–sphingomyelin ratio, fluorescence polarization). By analyzing fetal lung surfactant excreted into amniotic fluid, these tests evaluate the potential for normal breathing after birth.

Because of the relatively recent development of clinical laboratory analysis of amniotic fluid, relatively little information is available on proper handling of amniotic fluid specimens. Controlled studies to determine optimal methods of transport, storage, and centrifugation, among others, are rare. Most of the following information is therefore derived from published observations, informal studies, and extrapolation from practices of handling other body fluids.

TRANSPORT

Immediately after amniocentesis, the amniotic fluid should be transferred aseptically from the amniocentesis syringe into an appropriate sample container. Specimens for spectrophotometric analysis (for bilirubin) should be placed in an amber-colored container or otherwise protected from light (container wrapped in aluminum foil). Specimens for all types of culture and genetic studies should be placed into transport universal containers. Polystyrene is preferable to glass because the cells may adhere to glass.

The sample container should be wrapped in aluminum foil or placed in an opaque bag and delivered immediately to the laboratory. Specimens to be transported by mail or courier should be packed in an insulated container along with a coolant to maintain a low, but not freezing, temperature. If the specimen is for cell culture, the method of transporting fluid is important because the cells must remain viable. If transportation will take longer than 3 days, it may result in inadequate cell proliferation.

EXAMINATION OF THE SPECIMEN

Upon arrival in the laboratory, the specimen should be examined visually. Conditions that may interfere with analysis and that may be detected in this way include the presence of colored substances, blood, and turbidity.

INTERFERENCE WITH AMNIOTIC FLUID DETERMINATION—BILIRUBIN AND MECONIUM

Yellow color in amniotic fluid may indicate the presence of bilirubin but usually will not interfere with analysis. A greenish color is due to biliverdin and is the result of meconium, material from the fetus' lower gastrointestinal tract expelled when the fetus is distressed, being present in the fluid. The green color of meconium has an absorbance peak at 405 nm to 410 nm and interferes with the spectrophotometric determination of bilirubin at 450 nm, preventing an accurate analysis of bilirubin. In addition, meconium in amniotic fluid interferes with the determination of fetal lung maturity by fluorescence polarization.

BLOOD CELLS AND PLASMA

Blood of maternal origin is the most common contaminant of amniotic fluid and may interfere with amniotic fluid analysis. Spectral analysis for bilirubin cannot be performed if hemoglobin is present. If the red blood cells are not hemolysed, prompt centrifugation will remove them. Substances in blood plasma, however, may also interfere with analysis. Maternal plasma contains pigments that have absorbances at 450 nm. Small amounts of maternal blood (determined by the proportion of red blood cells present in the amniotic fluid sediment after centrifugation) will not affect the results significantly. Moderate amounts of maternal plasma in amniotic fluid may be compensated for by adding an appropriate amount of maternal plasma to the spectrophotometer blank, but gross presence of serum (proportion of erythrocytes to amniotic fluid greater than 0.005) prevents accurate bilirubin determination.

Fetal blood in amniotic fluid specimens will interfere because of the high bilirubin concentrations in fetal plasma. Fetal plasma cannot be used in a spectrophotometric blank; thus moderate amounts of fetal blood will prevent accurate bilirubin determinations. Whenever moderate amounts of blood are present in amniotic fluid specimens, the maternal or fetal origin must be determined by testing for fetal hemoglobin.

The presence of blood will interfere with determinations of fetal lung maturity because of the phospholipids in plasma. Only the determination of phosphatidylglycerol, not present in plasma, will give an accurate assessment of fetal lung status when amniotic fluid specimens are contaminated with blood. Blood contamination will also interfere with determination of amniotic fluid protein and other constituents present in plasma.

TURBIDITY

Turbidity of amniotic fluid interferes with spectrophotometric determination of absorbances determined directly on the fluid (as for bilirubin) or after a color reaction (as for total protein determination). The turbidity may be eliminated by centrifugation or filtration.

URINE

Occasionally a specimen thought to be amniotic fluid instead will be maternal urine. If a specimen brought to the laboratory is thought to be maternal urine, creatinine concentration should be determined. Amniotic fluid creatinine concentration is rarely greater than 4 mg/dl, whereas urine creatinine concentration is usually greater than 80 mg/dl.

CENTRIFUGATION

The cellular content of amniotic fluid forms a pellet at low centrifugal force; 140 $\times g$ is sufficient. Cells concentrated by centrifugation for examination with or without cell culture should be spun at low speed. Some authors recommend, however, a culture technique without centrifugation to increase the yield of cells.

For biochemical or spectral determinations of amniotic fluid, a centrifugal force of about 4000 $\times g$ for 10 minutes will remove cells, cellular debris, and most phospholipid lamellar bodies present in third trimester pregnancies. Higher forces, 7000 $\times g$ to 12000 $\times g$, may be used for specimens that remain turbid after lower centrifugal speeds have been used.

Centrifugation, even at low speeds (140 $\times g$), will remove some of the phospholipid lamellar bodies that contribute to measurements of fetal lung maturity. For this reason, when performing Clements' rapid foam stability test for surfactant (shake test), the specimen should not be centrifuged.

The literature conflicts on the effect of centrifugation on results of the lecithin–sphingomyelin (L–S) ratio. High speeds (10000 × g and higher) seem to decrease the L–S ratio, and lower speeds may also have an effect. A moderate speed, 3000 × g to 4000 × g, for 5 to 10 minutes should be used.

Fluorescence polarization determinations of lung maturity are stable over the centrifuged forces 1239 × g to 5000 × g. A force of 1500 × g for 10 minutes should be used.

SUGGESTED READING

1. Bernes EW, Crolla L: Determination of uric acid, creatinine, total protein and osmolality in amniotic fluid as a measure of fetal maturity. In Natelson S, Scommegna A, Epstein MB (eds): Amniotic Fluid: Physiology, Biochemistry and Clinical Chemistry. New York, John Wiley and Sons, 1974
2. Cherayil GD, Wilkinson EJ, Borbowf HI: Amniotic fluid lecithin/sphingomyelin ratio changes related to centrifugal force. Obstet Gynecol 50:682–687, 1977
3. Clements JA, Platzker ACG, Tierney DF, Hobel CJ, Creasy RK, Margolis AJ, Thibeault DW, Tooley WH, William OH: Assessment of the risk of the respiratory distress syndrome by a rapid test for surfactant in amniotic fluid. N Engl J Med 286:1077–1081, 1972
4. Dito WR, Patrick CW, Shelly J: Clinical Pathologic Correlations in Amniotic Fluid. Chicago, American Society of Clinical Pathologists, 1975
5. Fairweather DVI: Techniques and safety of amniocentesis. In Fairweather DVI, Eskes TKAB (eds): Amniotic Fluid—Research and Clinical Application, 2nd ed. Amsterdam, Excerpta Medica, 1978
6. Fex G, Holmberg N–G, Löfstrand T: Phospholipids and creatinine in amniotic fluid in relation to gestational age: I. Normal pregnancy. Acta Obstet Gynecol Scand 54:425–436, 1975
7. Fuchs F, Cederquist LL: Use of amniotic fluid cells in prenatal diagnosis In Fairweather DVI, Eskes TKAB (eds): Amniotic Fluid—Research and Clinical Application, 2nd ed. Amsterdam, Excerpta Medica, 1978
8. Gluck L, Kulovich MV, Borer RC: Diagnosis of the respiratory distress syndrome by amniocentesis. Am J Obstet Gynecol 100:440–445, 1971
9. Linback T, Frants T: Effect of centrifugation on amniotic fluid phospholipid recovery. Acta Obstet Gynecol 135:337–343, 1979
10. Oulton M: The role of centrifugation in the measurement of surfactant in amniotic fluid. Am J Obstet Gynecol 135:337–343, 1979
11. Pritchard JA, MacDonald PC: Williams Obstetrics, 15th ed. New York, Appleton-Century-Crofts, 1976
12. Simon NV, Elser RC, Livisky JS, Polk DT: Effect of centrifugation on fluorescence polarization of amniotic fluid. Clin Chem 27:930–932, 1981
13. Tegenkamp TR, Tegenkamp IE: Cytogenetic studies of amniotic fluid as a basis for genetic counseling. In Natelson S, Scommegna A, Epstein MB (eds): Amniotic Fluid: Physiology, Biochemistry and Clinical Chemistry. New York, John Wiley and Sons, 1974
14. Wagstoff TI, Whyley GA, Freedman G: Factors influencing the measurement of the lecithin/sphingomyelin ratio in amniotic fluid. J Obstet Gynecol Br Commonw 81:264–277, 1974

Part III

Handling Procedures

7

Transport Systems

Transport of Blood Specimens by Pneumatic Tube

James D. Jones

Clinical laboratories have been centralized in larger hospitals and in medical complexes to improve efficiency and cost-effectiveness. Centralization often entails delayed receipt of the specimen from the patient unless some mechanical conveyance is available for rapid delivery of specimens to the laboratory. The pneumatic tube offers an excellent solution to the problem. It is relatively trouble free, has virtually no restrictions on vertical or horizontal distances, and can be installed in a minimum of space. Many institutions have had favorable experience with such systems for transport of other materials such as records, drugs, x-rays, and laboratory reports. At the Mayo Clinic, pneumatic tubes have been used to transport records and other items since 1948.

HISTORY OF PNEUMATIC TUBES AND CLINICAL SPECIMEN TRANSPORT

The literature on application of pneumatic tubes to transport of specimens for clinical analysis is limited. McClellan and coworkers published a study in 1964 on the use of a pneumatic tube for transport of whole blood specimens.[1] They transported specimens in 15-ml Vacutainer tubes, filled by vacuum and placed into a rigid tube insert. Because of transport, mean values increased for serum potassium, from 4.5 to 4.9 mEq/L, hemoglobin, from 3 to 24 mg/dl, and lactic dehydrogenase, from 378 to 614 units. These increases exceed the allowable variation for clinical use. McClellan and associates also warned against the use of glass tubes because of possible breakage.

Delbrück and Poschmann used an experimental tube system in which they varied speed from 3 to 8 m/sec, and varied the number of switches and distance.[2] They reported that the increased potassium, lactic dehydrogenase, and hemoglobin levels were greater in serum transported in partially filled tubes of whole blood than in that transported in more completely filled tubes. They recommended transport of completely filled tubes at 3/msec. These authors used a rigid tube insert for the carrier.

At the Mayo Clinic in 1967, a desire to provide more complete laboratory service to its two hospital affiliations from a central laboratory 24 hours a day prompted an evaluation of the existing pneumatic tube system for transport of specimens for those tests available as emergency tests. Our data and experience were published in 1971.[3] At that time, we identified some restrictions on the use of the pneumatic tube for the specimen transport that were related to the size and weight of specimens, 24-hour urine specimens and centesis fluids, and the nature of the collected specimens for blood culture. In that evaluation, the effect of transport was determined for some analytes of interest to us and for some selected as indices of cellular destruction. Data obtained on serum sodium, chloride, carbon dioxide, calcium, creatinine, phosphorus, uric acid, bilirubin, total protein, whole blood urea, glucose and plasma fibrinogen demonstrated that changes owing to pneumatic tube transport of whole blood when a vinyl insert was used did not exceed the analytical variation. Although statistically significant, the increase that occurred in serum potassium concentration from whole blood specimens transported by pneumatic tube was considered clinically insignificant, the largest change observed being only 0.2 mEq/liter of serum.

The desire to provide a short turn-around time for serum enzymes prompted us to reevaluate the problem of cell destruction in the pneumatic tube and include determination of serum enzyme activities. Details of these investigations are presented later.

Our publication in 1971[3] was followed by one of Pragay and coworkers in 1974.[4] Using serum hemoglobin, potassium, and lactic dehydrogenase as indices of destruction, they concluded that the pneumatic tube could be used satisfactorily to transport blood specimens in either partially or completely filled tubes. Their rate of transit was 6.1 m/sec with a gradual deceleration. These authors used styrofoam inserts to hold specimen tubes. In a more recent publication, they confirmed their previous conclusions in an updated system and now appear to use "foam rubber padding" in the carrier.[5]

Pozanski and coworkers reported in 1978 that their low velocity, 3.6 m/sec, pneumatic tube was acceptable to transport whole blood specimens. In their system, which used foam rubber inserts, only partial filling of the Vacutainers did not affect the results adversely.[6] Serum lactic dehydrogenase activity was increased, however, if specimens were allowed to clot before transport.

Weaver and coworkers evaluated a computer-directed pneumatic tube system for transport of blood specimens.[7] They compared values obtained from specimens collected in evacuated tubes, then transported manually or by pneumatic tube. They experienced "no breakage, spillage, or leakage in any test."

Of 15 chemical and 6 hematologic procedures studied, only lactate dehydrogenase activity in the pneumatic tube transported specimens exceeded the analytical precision of the test.

A loss of some weak blood group antibodies was reported by Nosanchuk in 1977.[8] He concluded that pneumatic tube transport was satisfactory for hematology, coagulation, and urinalysis specimens. Although statistically different values owing to transport were observed for lactic dehydrogenase, both aspartate aminotransferase and glucose values obtained on transported specimens were acceptable for clinical use.

DESCRIPTION OF PNEUMATIC TUBE INSTALLATION AT THE MAYO CLINIC

The pneumatic tube system used in the Mayo complex was installed by Lamson Corporation. It is a unidirectional, continuously operating system not dedicated to laboratory use but also to transport mail, x-rays, and reports. The data published by us in 1971 were obtained from use of the tube between a chemistry laboratory and a hospital laboratory, a distance of 1423 meters that included 2 monitoring devices, 3 switches, and 67 bends (62 of 152-cm radius, 5 of 76-cm radius). The approximate speed of the carrier was 9.75 meters/sec.

In 1973, Amsco modified our system and installed a computer control, multiple loop twin exchanges, and soft landing. The speed of the carrier was reduced to 5.6 to 6.4 m/sec, and only 152-cm radius bends are now in the laboratory loop of the system. The computer allows monitoring of the system and immediate identification of stoppages.

Two shapes of Plexiglas carriers with a hinged leather flap at one end are used for laboratory samples. A 10-cm round unit is used within the hospital and the 10-cm oblong unit between the hospitals and laboratories (Fig. 7-1). The inserts are of vinyl lined with sponge rubber and are made to fit snugly within the Plexiglas carrier. A leather belt with loops sewn at a 60° angle is used to separate specimens within the 10-cm oblong carrier (Fig. 7-2).

Few restrictions on specimen container selection have been needed for transport through our present pneumatic tube. We do not send large volumes of urine or other biologic fluids through the tube because it is more convenient to sample and to aliquot those fluids centrally. The Mayo Clinic Blood Bank does routinely send units of cross-matched blood from a centrally located blood bank to both hospitals.

The plan for evaluating use of our tube system included assuring the reliability of mechanical performance and assessing the reliability of test results when samples were transported through the system.

1. Reliability and stability of equipment in mechanical performance.
   This aspect of the evaluation was extremely important at the Mayo Clinic because our tube system is not dedicated to the laboratory function only but is used for records, x-rays, and other items, increasing the probability of mechanical failures such as overloading switches. An unreliable system could cause prolonged delays and even loss of specimens if samples should be delivered to unmanned stations.

Fig. 7-1. *Carriers with inserts for transporting specimens in pneumatic tube system. Carrier on left is 10 cm (4 in) round; carrier on right is 10 cm (4 in) oblong.*

Fig. 7-2. *Inserts for carriers for transporting specimens in pneumatic tube. Screw-capped and Vacutainer tubes shown in inserts.*

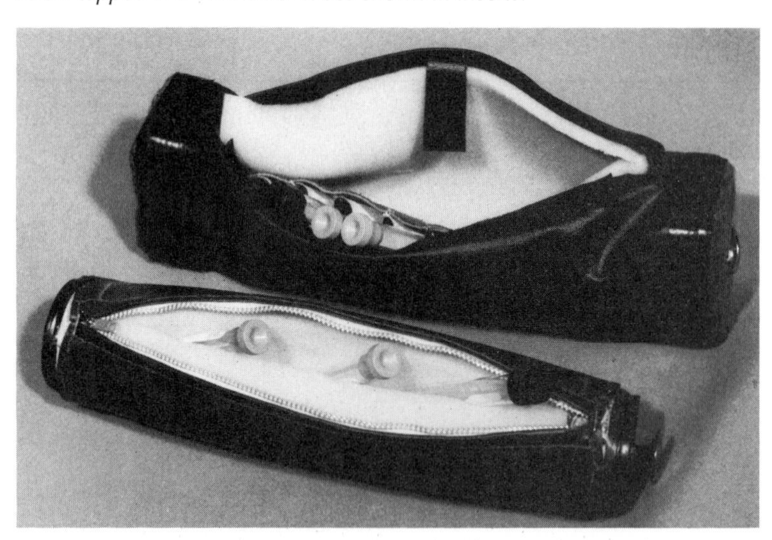

The introduction of computer-directed and -monitored control systems has improved the reliability of our pneumatic tube transport. Overload, changes in the rate of transport, or other mechanical alterations in the system have not caused problems in our experience. However, the nature of failures, type of service, personnel required, and maintenance costs are peculiar to an installation and must be explored by the user of each system.

2. Destruction of specimen vials, tubes, and containers *is a concern because of loss of specimen and contamination of the transport system. In our experience, Vacutainers have been used for most samples. Very seldom have we experienced a cracked or damaged Vacutainer. Other containers are always checked for sturdiness before use.*

PROCEDURES USED IN ASSESSING RELIABILITY OF TEST RESULTS

The analytical methods used were the standard procedures currently used in the clinical laboratories. The blood constituents chosen for the study were those available at the Mayo Clinic for medical emergencies plus serum hemoglobin and lactic dehydrogenase, which were included as indices of cell destruction.

Reproducibility of the methods was established from data obtained on ten aliquots of blood taken into 10-ml Vacutainers from one person at one venipuncture without stasis. After standing 30 minutes, the tubes were rimmed with wooden applicator stick and centrifuged to obtain serum for analysis. For other studies, multiple blood samples were obtained by venipuncture from patients and laboratory personnel by syringe, and then aliquots were delivered into tubes as indicated.

The venipunctures and analyses were performed by a technologist. Centrifuging and analysis of specimens were performed in a batch to reduce variables—for example, time of serum contact with cells, temperature, and drift in analytical procedure.

STABILITY OF THE SAMPLE

A concern in the transport of whole blood is the destruction of formed elements, especially of red blood cells. The ratio of erythrocyte-to-plasma concentrations for potassium is 100:4.4 mEq/liter and for lactic dehydrogenase, 58,000:360 units/liter.[9] One would predict by calculation that 1% hemolysis would cause a true plasma potassium value of 3.5 to increase to 4 mEq/L (+14%), and similarly for lactic dehydrogenase to increase from 360 to 650 units (+80%). Hemoglobin at 15 g/dl of whole blood and <10 mg/dl of plasma is another analyte with a large cell-to-plasma gradient.

Our initial studies had shown that the considerable frothing and marked increases in serum hemoglobin, potassium, and lactic dehydrogenase values that occurred in specimens transported in vials in a rigid tube insert could be reduced by use of a soft sponge, vinyl-covered insert. Later inclusion of serum enzymes in the group of tests available on an emergency basis that required a turn-around time of less than 1 hour led us to consider further precautions to assure that red blood cells remain intact during tube transport from the hospitals. Two approaches were considered: separation of cells from serum before transport; or evaluation of the effectiveness of completely filling the specimen tubes to reduce the variability in determining lactic dehydrogenase activity.

Seven aliquots of whole blood from a venipuncture with a syringe and without stasis were placed in plain 5-ml Vacutainer tubes. The value for precision of the standard technique for lactic dehydrogenase was a mean of 80 units/

Table 7-1. *Variability of LDH Values*

TUBE	LDH
	units/liter
1	78
2	75
3	78
4	81
5	83
6	83
7	81
$\bar{X}$	80 (75–83)

Table 7-2. *Effect of Transport of Whole Blood on Serum LDH Activity*

	VACUTAINER		SCREW-CAPPED GLASS	
SPECIMEN	*Control*	*Transport*	*Control*	*Transport*
		units/liter		
1	112	155	112	114
2	77	135	79	92
3	90	116	85	89
4	96	119	95	95
5	94	121	89	93
6	82	95	78	80
7	83	114	84	82
8	97	130	102	100
9	75	124	76	78
10	74	97	73	76
11	74	109	74	75
$\bar{X}$	87.0	119.5	86.1	88.5

liter, with a range of 75 to 83 units/liter (Table 7-1). Four aliquots of blood from each of 11 persons were obtained without stasis. Two were placed in Vacutainers and two in screw-capped vials (tube No. 99447, Corning Glass Works). One of each type container was transported round trip before concurrent separation from cells and analysis with the matching nontransported specimens. The data are shown in Table 7-2. The variation between control specimens for each individual is similar to that observed within an individual, giving additional data on the precision of the lactic dehydrogenase determination for the entire technique.

Lactic dehydrogenase activity in the specimens transported in Vacutainers were all increased. The average values increased from 87 to 119.5 units/liter, with increases in individual samples from +13 to +58 units/liter. The aliquots transported in filled screw-capped glass vials exhibited minimal changes; the means increased from 86.1 to 88.5, with a range of differences of −2 to +13 units/liter. This change due to transport is an absolute, not a percentage, change and has been considered acceptable in the emergency situation. It has not been exceeded in specimens from patients with blood dyscrasias or elevated lactic dehydrogenase activities.

Creatinine kinase activity in the same aliquots gave similar values for control Vacutainer, control screw-capped, and transported screw-capped—35.3, 35.2, and 35.6 units/liter, respectively. The largest differences between aliquots were 2 units/liter, well within the precision of the method.

Serum hemoglobin concentrations in these specimens were also unaffected by transport in screw-capped vials. Mean hemoglobin values for Vacutainer nontransported control, screw-capped control, and screw-capped transported were 2.9, 2.7 and 2.6 mg/dl, respectively; the range of the differences in the screw-capped control comparison was −2.3 to +2.3 mg/dl, indicating that the differences apparently were due to factors other than hemolysis from transport.

The bias for serum potassium from specimens of whole blood transported in the Vacutainer was +0.12 mEq/liter of serum, similar to data presented earlier. The largest increase due to transport in screw-capped vials was 0.1 mEq/liter and occurred in 4 of 11 specimens in this series. This increase, although appearing as a plus bias, is within the precision of the method and has been considered acceptable.

Denaturation of proteins, which would alter the electrophoretic, catalytic, or immunologic properties, may be caused at the air–liquid interface in a shaken tube. We have not been able to demonstrate any change in protein properties in specimens transported in our system. Neither has a study with the antibodies used in cross-matching blood demonstrated any alteration in protein stability (Taswell H: Personal communication). A recently published report has indicated that many serum and plasma constituents are remarkably stable during shaking.[10]

We have not been able to demonstrate an increase in temperatures of specimens transported by pneumatic tube. Specimens in tubes that require ice, such as blood gas, are placed in plastic bags that contain ice and are maintained easily at cold temperatures during delivery.

CURRENT USE AT THE MAYO CLINIC

The data presented above showed that we could centralize our laboratories. Since 1967, we have relied on the pneumatic tube as the primary specimen delivery system for the emergency laboratory services provided to two large hospitals, 1000 and 800 beds, from a centralized laboratory that also serves outpatients. Specimens for all emergency chemistry, hematology, and blood banking activities are currently transported by pneumatic tube to the central facility. Laboratory test reports are returned to the point of origin, hospital or clinic, by the pneumatic tube.

We have found no limitations except for size and weight of specimens on the use of the system for transport of clinical specimens. A major concern is the overuse by others for nonlaboratory purposes that may overburden the switching devices in our system, resulting in slower transit time.

Data have shown that complete filling of screw-capped glass tubes is needed to prevent destruction of cellular components, as shown by alterations in serum hemoglobin and lactic dehydrogenase activity.

Spurious hyperkalemia caused by loss of potassium by formed elements in specimens from patients with thrombocythemia[11] or high white blood cell counts without increased platelet counts, as in chronic lymphocytic leukemia,[12] occurs independently of transport of the whole blood specimens. True values for plasma potassium for such specimens are obtained only by separating plasma from formed elements immediately after obtaining the specimens by venipuncture.

Our experience and a review of the limited literature available on the documented use of a pneumatic tube system for transport of blood specimens indicates that the system can be used effectively for transport of specimens for most analyses; there are minor differences in the data from the different installations; and each installation should be evaluated individually before unlimited use of transport specimens to the laboratory.

RECOMMENDED ACTION FOR EVALUATING PNEUMATIC TUBE TRANSPORT OF CLINICAL SPECIMENS

1. Assure mechanical reliability of the system.
2. Check for destruction of cellular components by determining if serum potassium or lactic dehydrogenase is increased in specimens from normal persons by transport by the tube. Comparison of split specimens requires elimination of variables introduced in venipuncture and analysis (*see* Chap. 10).
3. Do not assume that stability of one analyte indicates stability of others. Use round trip or two times any expected distance, including switches, for testing the acceptability before extending the use to a new analyte.
4. Transport split specimens—one by pneumatic tube and one by surface or hand carried from patients with disease—and compare values obtained for the analytes of interest. This exercise will also allow an evaluation of the entire system operated by "users" under routine conditions.

REFERENCES

1. McClellan EK et al: Effect of pneumatic tube transport system on the validity of determinations in blood chemistry. Am J Clin Pathol 42:152, 1964
2. Delbrück VA et al: Über den Einfluss des Rohropostransportes auf klinisches Untersuchungsmaterial unter verschiedenen Betriegsbed ingungen. Z Klin Chem Klin Biochem 6:28, 1968
3. Steige H et al: Evaluation of pneumatic-tube system for delivery of blood specimens. Clin Chem 17:1160, 1971
4. Pragay DA et al: Evaluation of an improved pneumatic-tube system suitable for transportation of blood specimens. Clin Chem 20:57, 1974
5. Pragay DA et al: A computer directed pneumatic tube system: Its effects on specimens. Clin Biochem 13:258, 1980
6. Poznanski W et al: Implementation of pneumatic-tube system for transport of blood specimens. Am J Clin Pathol 70:291, 1978
7. Weaver DK et al: Evaluation of a computer-directed pneumatic-tube system for pneumatic transport of blood specimens. Am J Clin Pathol 70:400, 1978
8. Nosanchuk JS: Automated transport of clinical laboratory specimens by a new air-transport tube system. Am J Clin Pathol 67:204, 1977
9. Caraway WT: Chemical and diagnostic specificity of laboratory tests. Am J Clin Pathol 37:445, 1962

10. Felding P et al: The stability of blood, plasma and serum constituents during simulated transport. Scand J Clin Lab Invest 41:35, 1981
11. Whitfield JB: Spurious hyperkalaemia and hyponatraemia in a patient with thrombocythaemia. J Clin Pathol 19:496, 1966
12. Bellevue R et al: Pseudohyperkalemia and extreme leukocytosis. J Lab Clin Med 85:660, 1975

Transport Equipment

Mary Mein

Sending specimens from outlying patient areas to a centralized laboratory in specially designed containers can be economical and efficient. Reusable and prelabeled containers can ensure safe transport of blood and fluids and, with microbiology samples, ensure safe transport of critical specimens.

SPECIMEN BLOCKS

Plastic blocks with 6.25 cm (2.5 in) circular openings have proved to be safe and efficient for transporting evacuated tubes that contain blood (Fig. 7-3).

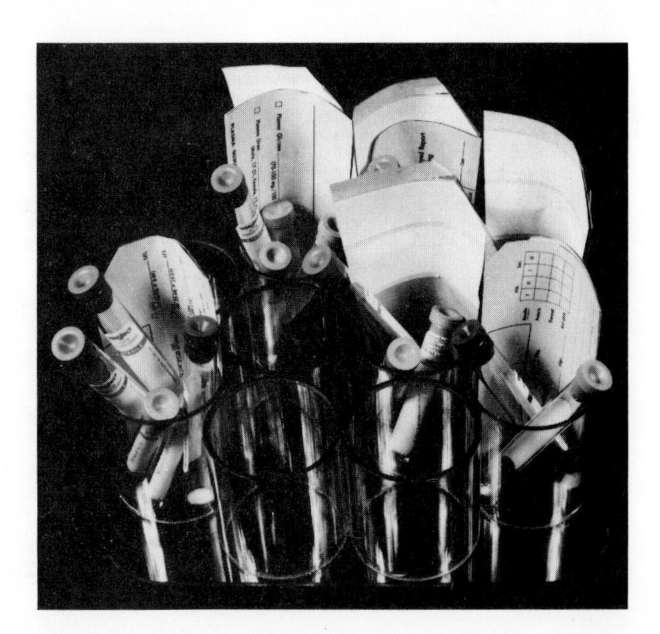

Fig. 7-3. *Specimen block, 20 × 25 cm (8 × 10 in), 7.5 cm (3 in) high, 5 cm (2.5 in) diameter holes.*

These blocks can be used on the venipuncturist's cart. The forms, labels, and blood samples from each patient can fit in one of the circular openings. The entire block can be placed in a transport box to be sent to the central laboratory area, reducing the handling of specimens and forms. Having all of one patient's samples together speeds sorting in the central processing area.

TRANSPORT BOXES

Transport boxes of high impact plastic (Fig. 7-4) can be purchased in many sizes. Canvas straps can be fitted to them to facilitate carrying, and Velcro strips can be added to keep the covers in place. Marking pens should be used to address destination and return. These boxes and blocks stack well for storage. Specimens are rarely left in this type of transport box because messengers or laboratory workers can see at a glance if the container is empty. The blocks and boxes are washable and autoclavable

Fig. 7-4. *Transport boxes.* (Left) *Large, 25 × 33 cm (10 × 13 in), 15 cm (6 in) high.* (Right) *Small, 15 × 18 cm (6 × 7 in), 15 cm (6 in) high.*

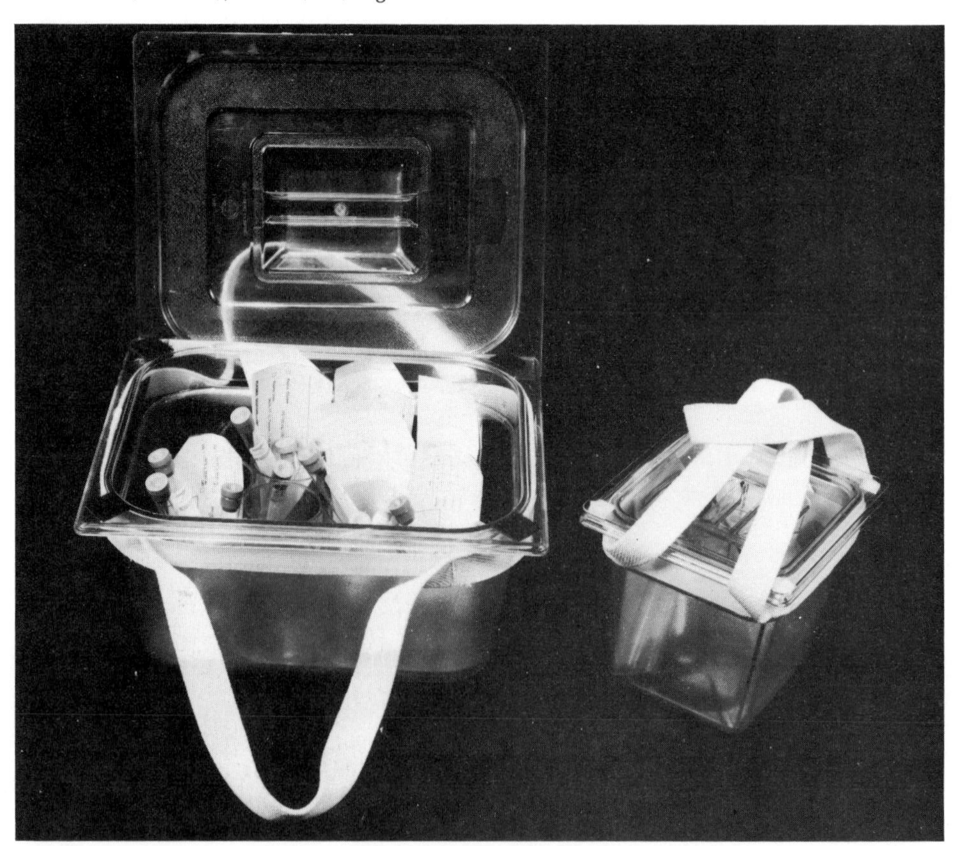

URINE SPECIMENS

SINGLE URINE SPECIMENS

Boxes and blocks can also be used for transporting single sample urine specimens; however, their use for stool specimens have proved unsatisfactory because they become odoriferous and are offensive to staff and messengers.

24-HOUR URINE SPECIMENS

Twenty-four hour urine specimens should be collected in 3-liter large plastic bottles that will hold the whole collection. Fractional collections can be made in smaller bottles with good leak-proof screw caps. These containers (Fig. 7-5) should be transported in a plastic draw string bag.

REFRIGERATED SPECIMENS

Certain specimens for microbiologic study such as urine and sputum cultures should be transported in a container that will ensure a temperature of between 2°C and 10°C. Steel-lined insulated boxes can be used (Fig 7-6). These boxes are

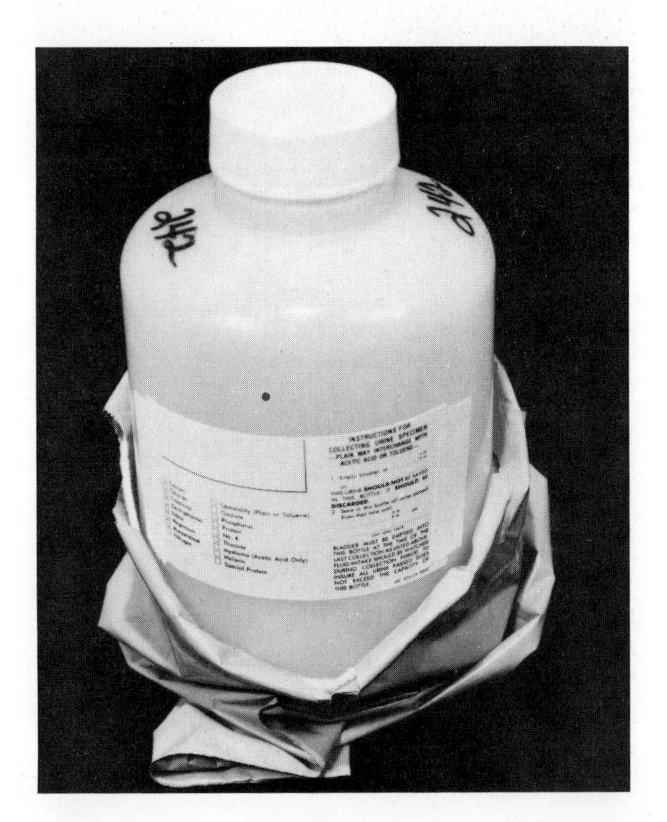

Fig. 7-5. *Urine containers.*

Fig. 7-6. *Steel-lined insulated box.*

Fig. 7-7. *Metal rack with blood cultures and other microbiologic samples.*

kept refrigerated until used to ensure that the specimen will maintain a temperature of 2°C to 10°C during transportation. An alternative to having such boxes constructed would be to use a styrofoam liner for the transparent transport boxes with a commercial coolant to control the temperature. Blood cultures and other microbiology samples that can be left at room temperature can be placed in metal racks (Fig. 7-7) that fit into the plastic transport previously described.

STOOL SPECIMENS

Individual stool specimens in cardboard containers can be placed in polyethylene bags, sealed by knotting the top of the bag, and then placed in addressed paper bags.

DESIGN

Permanent transport containers are cost efficient after the initial purchase. They should be designed to fit an institution's specific needs. Containers that are transparent and autoclavable are preferable.

One should consider the following things when designing a specimen transport container.

1. Temperature control needs
2. Protection of specimens to control breakage and spillage
3. Sturdiness
4. Washing and autoclaving potential
5. Cost
6. Appearance

External Transport of Medical Specimens

Gerald Wollner

The field of clinical laboratory medicine has changed significantly in scope, size, and technical complexity during the past 20 years. Government regulations and cost constraints more recently have had an impact on this medical discipline. One of the results has been an increased number of specimens being sent to reference laboratories.

Transporting specimens to other laboratories requires proper attention to the specifications for the collection, handling, and shipment of this human material. These criteria contribute to the final result derived from a laboratory test.

Sending specimens to a regional or commercial laboratory requires not only the usual precautions, such as proper identification, appropriate preservative, and patient preparation, but also attention to adequate packaging, labeling, and shipping requirements. My objective in this section is to provide a practical and systematic method for maintaining specimens during transport in a satisfactory biologic condition for subsequent analysis. This objective was formulated on the premise that tests performed on diagnostic specimens are worthless if the specimen is received in an unsuitable condition. In most instances, specimens sent to another laboratory for testing are not urgent medically. Also, some time will elapse between procuring the specimen and its receipt by the reference laboratory. During that time, environmental conditions and certain biologic or chemical transformations, or both, may affect the stability or suitability of the specimen. This section presents minimum guidelines of acceptable practice and methods for handling the transported medical specimens.

REGULATIONS

DIAGNOSTIC SPECIMENS

Diagnostic specimens are defined in the Public Health Service Interstate Quarantine regulations 42 CRF, Section 72.25, as any human or animal material including, but not limited to, excreta, secreta, blood and its components, tissue, and tissue fluids being shipped for diagnosis. Diagnostic specimens in themselves are not subject to any regulations. From a practical safety standpoint, however, it is advisable to handle all diagnostic specimens in a way that protects those persons or materials that come into contact with them.

ETIOLOGIC AGENTS

Etiologic agents, specimens that contain viable microorganisms, are regulated by the Interstate Quarantine regulations. Specimens that contain viable microorganisms are subject to special packaging and labeling requirements that will be reviewed later in this section. These requirements are set forth to guard against the leakage of the contents under normal transportation conditions and to alert persons handling these packages as to potential danger.

The list of microorganisms governed by this regulation was first published in the *Federal Register* on June 20, 1972 and later revised in the same publication on July 21, 1980. This list may be revised from time-to-time by notice published in the *Federal Register.* For example, a recent addition to this list is the hepatitis-associated antigen.

A selected number of etiologic agents must be sent by registered mail or an equivalent system that requires or provides for sending notification of receipt to the sender immediately upon delivery. A current list of etiologic agents is available from the Office of Biosafety, Center for Disease Control, 1600 Clifton Road, N.E., Atlanta, GA 30333.

DRY ICE

Another regulation that affects the transport of specimens is the Department of Transportation's Restricted Article Regulations. Dry ice is considered a hazardous material under these regulations. Hazardous materials usually require a Shipper's Certification for Restricted Articles form to accompany such materials. However, a medical specimen being refrigerated with Dry Ice is accepted from the shipping paper without this special form, thus being an exception. Labeling requirements for Dry Ice will be reviewed later in this chapter.

Occasionally, a package that contains Dry Ice for refrigerating medical specimens, even though appropriately labeled, will be rejected by an airline. The pilot has the ultimate authority over the cargo being carried, and he can refuse to accept the package. If he rejects a package, it usually reflects a misinterpretation or lack of knowledge of the exception in the Restricted Articles regulations for medical specimens shipped on Dry Ice.

TEST SPECIFICATIONS

The reference laboratory where the specimen is being sent should provide the necessary information and instructions on how to collect and handle the specimen. The sending laboratory must follow these instructions, contact the reference laboratory to see if they are able to accept the specimen under the circumstances. This communication will alleviate delays in test reporting and also will minimize misunderstandings or communication breakdowns.

If the reference laboratory provides specimen containers, request forms, mailers, and other materials, it is advisable to use them because most reference laboratories expedite requests arriving in their own packaging materials. Test requests written on a foreign form are handled as exceptions or problems, thus delaying their accessioning process.

Filing systems, internal specimen transfer systems, and other processes are also designed around the reference laboratory's provided materials. Adherence to instructions for specimen requirement, patient preparations, forms, and containers can save up to 24 hours in the return of a test result.

SPECIMEN IDENTIFICATION

Specimen identification, a long-standing problem in laboratory medicine, is intensified when specimens from one laboratory are sent to another. Each specimen, whether shipped by itself or with others as a multiple shipment, must be identified individually with appropriate patient name, reference numbers, or other identifiers. Specimen information records and laboratory forms should be enclosed in an envelope attached to the specimen, or placed within the shipping carton.

If a specimen is shipped on ice, dry or wet, the identifing label must be affixed securely to the specimen container. Cold, wet conditions create problems in keepng the identifying labels affixed securely. Additional taping is recommended to prevent this problem.

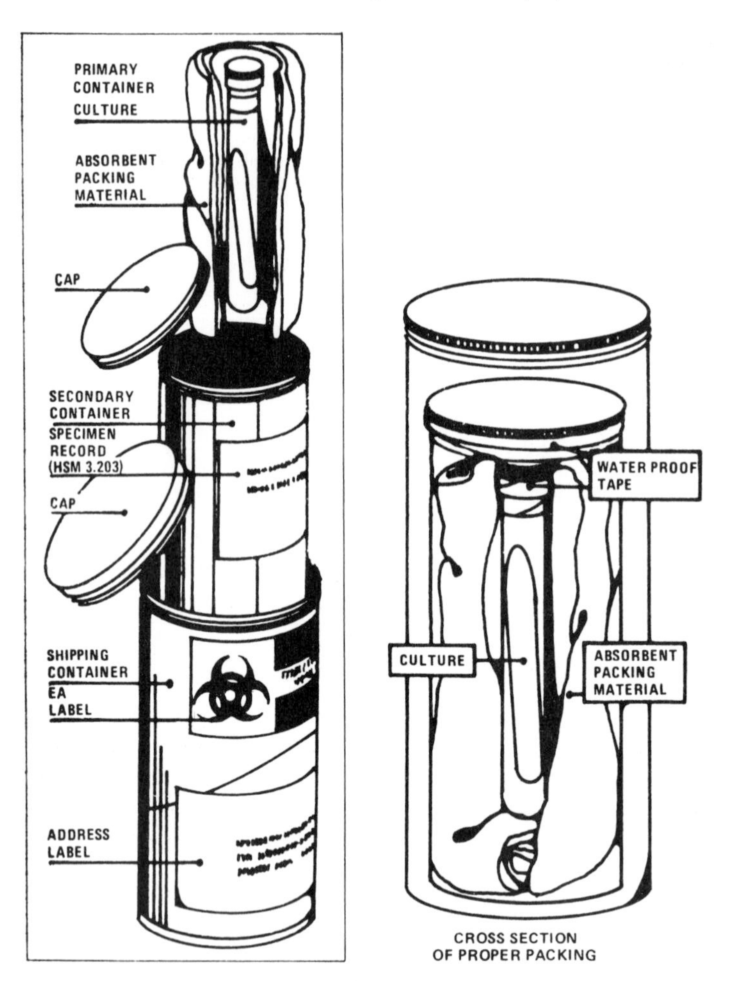

Fig. 7-8. *Packaging and labeling of etiologic agents.*

SPECIMEN CONTAINERS

The federal regulations cited previously describe the following characteristics for such containters and their shipping carton: "Must withstand leakage of contents, shocks, pressure changes and other conditions incident to ordinary handling in transportation."

PRIMARY CONTAINER

A container for transporting a specimen, commonly referred to as a "primary container," should be designed and constructed so that if subjected to adverse environmental conditions, its contents are not released to the environment and the effectiveness of the packaging is not impaired.

1. **Do Not** send more than 50 ml in one package.
2. This is **Not** to be used for international shipment.
3. Refer to separate sheet for instructions or proper packaging of etiologic agents.

This is to certify that the contents of this consignment are properly classified, described by proper shipping name and are packed, marked and labelled and are in proper condition for carriage by air according to all applicable carrier and government regulations. This consignment is within the limitations prescribed for passenger aircraft.

Numbers of package (s)	Specify each article separately (Proper shipping name)	Classification	Net quantity per package
	Etiologic Agent, n.o.s.	Etiol. Ag.	/ ml

Shipper:

Date ____________

____
(Signature of Shipper)

Fig. 7-9. *A notice to carrier.*

SPECIAL REQUIREMENTS

With etiologic agents, the primary container and its transport package must meet specific requirements (Fig. 7-8). The regulations include taping the stopper on the primary container, wrapping the container in absorbent material, and placing it in a secondary container that is crushproof and leakproof. A package that contains more than 50 ml of an etiologic agent has some additional packaging requirements specified in 42 CRF, Part 72.25.

In addition to the packaging requirements, a Notice to Carrier (Fig. 7-9) must be affixed to the package when it contains an etiologic agent.

Polypropylene and polyethylene containers are suitable for applications. Polystyrene containers, which are subject to cracking when frozen, are not recommended. Glass vials are also not recommended unless precautions are taken to prevent breakage.

Specimen containers must be leakproof, that is, they should be equipped with a stopper or a cap designed specifically for the container. All components of the container that come into contact with the specimen should be without any contaminating material that could alter the specimen chemically during storage for transportaton.

The specimen container should be examined for manufacturing defects. The top of the container should have a smooth surface on which to set the cap or stopper. The container opening should be symmetrical and not "out of round." The cap or stopper should be tested on the container before actual use to ensure proper fit.

After the specimen is transferred to the container, the cap or stopper should be secured tightly. Mechanical devices must not be used for further tightening because the container, cap, or stopper may crack or bend.

The reference laboratory at the Mayo Clinic, Mayo Medical Laboratories, devised a method for testing the leakproof characteristics of specimen containers. The test method subjects the vial to negative pressure and vibration, simulating an airplane cargo compartment. Although it is not practical for every laboratory to do such testing, the reference laboratory should be responsible for providing vials that have been tested for leakage.

SHIPPING CARTONS

The shipping carton for specimens must be able to withstand shocks and weight pressures commonly associated with handling during transportation. Single specimens can be shipped satisfactorily by enclosing the sealed primary container in a durable outer shipping carton. When multiple specimens are shipped in a single shipping carton, each primary specimen container must be protected individually to reduce shock and to prevent vial-to-vial contact.

Various carton configurations and materials are suitable for specimen shipping cartons. Padded shipping envelopes usually are adequate for ordinary transport purposes. Corrugated, fiberbound, styrofoam boxes or other materials with rigid characteristics are suitable if they are designed to fit the specimen container securely.

If refrigerated specimens are shipped, styrofoam or other materials with similar insulating qualities are most suitable. Such cartons for shipping Dry Ice must be vented to allow the escape of gases formed.

The size and type of the shipping carton along with the amount of refrigerant used are critical in maintaining the desired transport environment. If a frozen specimen is being shipped on Dry Ice and must be maintained for 48 hours, the following characteristics would be needed: a styrofoam carton with 2.5 cm (1 in) sidewalls, an interior capacity of 125 cubic inches, an ambient temperature of 22°C (72°F), and a solid piece of Dry Ice. Any variation to the carton wall thickness, the weight and texture of the Dry Ice, or the interior box dimensions would affect the refrigeration characteristics of the shipping environment.

LABELING

There are no specific requirements or regulations for labeling a diagnostic specimen unless it contains an etiologic agent or it is being shipped on Dry Ice. Unnecessary labeling may confuse persons coming in contact with the shipping carton. Packages marked "urgent"—or "medical specimen" are not given priority handling. Often such unnecessary labeling raises questions and exception handling procedures that may delay the shipment.

The specimen that contains a suspected or known etiologic agent must be sent according to the special packaging requirements reviewed above. In addition, the etiologic agent label shown in Figure 7-10 must be affixed to the shipping carton. When specimens are being shipped on dry ice, a hazardous material as defined in the Air Transport Restricted Articles Circular No. 6-D, the shipping carton must be labeled as indicated in Figure 7-11.

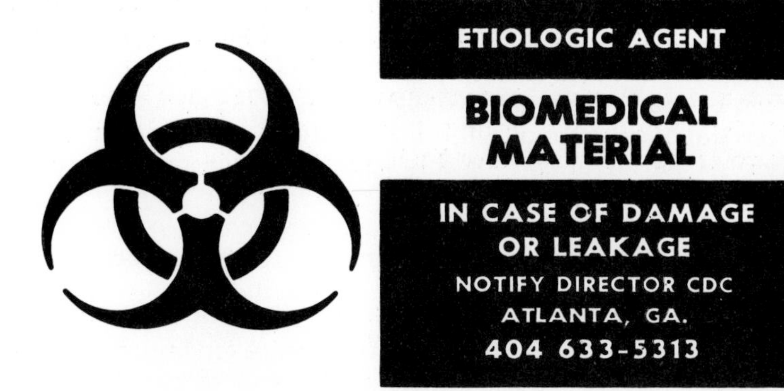

Fig. 7-10. *Etiologic agent label.*

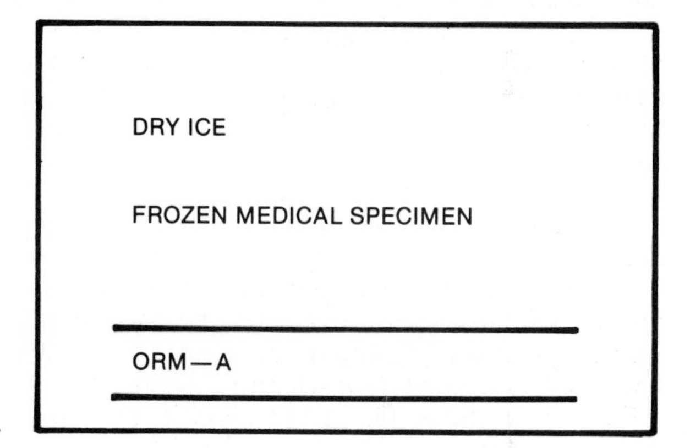

Fig. 7-11. *Shipping carton label.*

TRANSPORTING

The transportation industry, and innovative systems to capitalize on it, have been very instrumental in speeding medical specimens from one laboratory to another. Various modes of transportation are available for specimen transport, each with advantages and disadvantages in terms of cost, time, and convenience. This time-cost-convenience relationship is important in selecting where to send a specimen and also in determining whether its stability can be maintained. Usually, the faster and more personalized the service, the higher the cost.

Basically four types of transport services usually are considered for transporting medical specimens: U.S. Postal Service, air freight, courier services, and surface freight carriers. The U.S. Postal Service offers many different grades of service, but Priority Mail and Express Mail generally are the services of choice. The Priority Mail service selects the fastest routing possible, whether by air or by surface transport, to get a parcel from one location to another. The transit time is determined by distance and transportation systems available to the Postal Service. Thus, it is sometimes faster to mail a parcel from New York City to

Los Angeles than from New York City to a small community in upstate New York. Priority Mail timetables are available from the local post office.

Express Mail is a highly personalized service that guarantees delivery within 24 hours. Pick-up and delivery service are options. Express Mail is limited to certain cities, although the list of cities is growing, and it is an expensive service. One advantage of this service is that the quality is quite uniform throughout the country.

Small package air-freight services are available from most medium and larger cities in the country. This service can be from airport-to-airport or door-to-door depending on the service available in the community. Its cost is similar to that of Express Mail.

Courier service associated with reference laboratories has increased significantly in recent years. The principal advantage of courier service is the personal and convenience characteristics. The courier service driver usually carries Dry Ice and other refrigerant materials and picks up the specimen directly from the laboratory. Courier services are expensive, but they are gaining popularity in the clinical laboratory field.

Other surface transport services, such as bus companies and Amtrak, are also available for small parcels. These services require the parcel to be delivered and picked up, but there is some advantage to minimizing the number of persons who handle the parcel.

RECEIPT OF SPECIMENS AFTER TRANSPORT

If the reference laboratory to which specimens are sent for analysis does not have appropriate standards to measure the acceptability of a specimen, the laboratory work and the resulting data might be worthless, yet very costly to patients. The setting of high standards for laboratory practice, including specimen acceptance, usually separates the reference laboratories interested in good patient care from those concerned primarily with the commercial aspects of laboratory medicine.

The receiving laboratory should use the following control procedures to ensure that the integrity of the specimen was not compromised during initial handling, storage, and transport.

1. The specimen container must be identified with an appropriate control or accession number. If, however, the specimen is labeled only with the patient's name or the patient's identification number, the specimen may be processed if this information matches that on the request form.
2. If a specimen is labeled with the patient's name, along with a control or accession number, the name and number on the specimen should match the name and number on the accompanying request form.
3. Both the specimen and the shipper's specifications for transporting the specimen should be checked to ensure compliance.
4. The procedure for each test requested should be checked to determine that a sufficient amount of specimen has been received and that the specimen is appropriate for the tests requested. In many cases, the differentiation between serum and plasma is not readily apparent; therefore, information on

specimen type should be written on the specimen container label or the request form.

5. Notation of patient diet restrictions, if required for the test requested, should be included on the specimen container label or the request form.
6. Any additional patient history and clinical information required by the laboratory performing the test must accompany the specimen and the test request form.
7. If the specimen is hemolyzed, the test procedure should be checked to determine if hemolysis affects the test.
8. If the specimen is aliquoted within the receiving laboratory, the aliquot must be labeled and identified appropriately.

SUGGESTED READING

1. Air Transport Restricted Articles Circular No. 6-D, Section III, Part K, 615(e), p 240. Airline Tariff Publishing Company, USA, 20 June 1977
2. Code of Federal Regulations, 49 Transportation, Section 173.386, 173.387, and 173.388
3. National Committee for Clinical Laboratory Standards (NCCLS): Standard Procedures for the Handlng and Transport of Adopted Standard H4-A1. Domestic Diagnostic Specimens and Etiologic Agents. Villanova, Pennsylvania, NCCLS, 1979
4. Public Health Service Interstate Quarantine Regulations 42 CFR, Section 72.25, Etiologic Agents. Atlanta, Center for Disease Control, 31 July 1972
5. U.S. Postal Service Regulations (Domestic) TL-34, 2-7-75, Issue 97
6. Wilding P, Zilva JF, Wilde CE: Transport of specimens for clinical chemistry analysis. *Annu Clin Biochem* 14:301–306
7. Wollner GC: Handling, storing and transporting diagnostic specimens. *Lab Med* 2(2):87–91, 1979

8 Processing of Specimens

Fuad K. Mansour

The processing of laboratory specimens has gained greater attention as high technology and increased demands for quality control have evolved within the clinical laboratory. As the importance of specimen processing in generating meaningful laboratory results became more apparent, laboratories began searching for alternatives to the traditional methods. At the core of this search was the question, Should specimen processing take place in individual laboratories (the traditional manner), or would our purposes be served better by a centralized system of specimen processing?

There is no straightforward answer to this question. The solution depends on the size of the laboratory, the nature of the services, the financial and organizational constraints the laboratory operates within, and, perhaps, the individual preferences of those who make decisions regarding the laboratory.

This chapter discusses centralized processsing of laboratory specimens. Subjects include the physical layout, equipment, personnel, policies, and procedures required to make this concept work. Also included is a discussion of the advantages realized by Central Processing and the ways by which high technology issues and quality control may be addressed. The review of this material should provide an objective basis on which the decision of centralized versus distributed specimen processing can be made.

The accurate processing of specimens is very important if the results of the tests are to be meaningful. It is the duty of everyone who handles a specimen, from the time of collection to the time of testing, to preserve the chemical integrity of the specimen, that is, not to introduce contamination or in any other way to invalidate the patient's test result.

SYSTEMS OF PROCESSING

- *Individual Processing by Each Laboratory* The most common systems of specimen processing. After collection, individual specimens are sent to each laboratory to be processed and analyzed.

- *Centralization of Specimen Processing.* A new, unique system that recently has been introduced in several large institutions and clinics but may be applied to any sized laboratory.
  The rest of this section pertains to the concept of centralized processing; however, the procedures involved are beneficial to all involved in processing specimens.

OBJECTIVES OF CENTRAL PROCESSING AREA

1. Central control of specimen flow, allowing implementation of unified accessioning system for incoming specimens. *A central processing area acts as a funnel into which most specimens flow before distribution to individual laboratories. It is possible to implement an accessioning system with centralized control.*
   A unified accessioning system is needed for positive specimen identification. Such a system can be controlled centrally and has no potential of assigning duplicate accession numbers to different specimens because the system has control over the accession number assigned.
2. The ability to perform several pretesting functions on specimens before being forwarded to individual laboratories for analysis. *The technicians in the area are able to perform the following functions.*

   Generate aliquot labels
   Sort specimens into type of processing
   Place specimens into appropriate centrifuge
   Operate centrifuge
   Unload centrifuge
   Aliquot master specimen into appropriate tubes
   Place aliquot tubes into racks for delivery to the final laboratory

This processing allows the laboratory technologist more time to devote to the analysis of the test.

3. The ability to develop a "stored serum" program. *If a portion of the master specimen (in a refrigerated specimen bank) can be saved for a period of time, the laboratories possibly could obtain additional aliquots for retesting or additional determinations requested by the patient's physician. In general, patients do not like to give blood for testing, and it is upsetting and worrisome to them if they are called back for a redraw. By using the stored serum concept, the number of patient call-backs for redraws or for requesting additional determinations can be reduced.*
   When using this stored serum concept, each clinic or laboratory should determine which tests can be performed from stored serum. One should look at the methods used for determination, the period of stability of the specimens, and the temperature effect on the specimens.
   At the Mayo Clinic, the use of stored serum was established in 1974. In the procedure guide for physicians, the stored serum instructions and list of tests are given. The instructions are as follows.

Serum retained from Clinic and hospital patients age five and above, who have had a chemistry group is available for additional tests. This service permits the performance of some additional tests without requiring a second venipuncture. Two and one half to three ml of serum will be retained and refrigerated for seven days at Central Processing.[1]

At our institution the ten most common tests ordered by physicians from stored serum are the alpha-1 antitrypsin, free and total thyroxine, TSH, total thyroxine, alkaline phosphatase isoenzymes, cholesterol, B12, GGT, folate, and rheumatoid factor assay. In 1981, our physicians requested 4466 additional tests from stored serum, and the laboratories requested 18,694 specimens, a savings of 23,160 venipunctures.

4. To furnish a base for inquiry into the status of test requests. *Technologists from different laboratories can make one telephone call to get all the information needed about the patient's tests and where to find more serum. The physician is able to inquire if the specimen is drawn and is being delivered to the laboratory.*
5. To decrease the quantity of blood drawn from the patient for laboratory testing. *Another advantage of centralization is the decrease in the quantity of blood needed for laboratory tests. The concept of combination drawing is applied to lessen the volume of blood needed. With a combination draw, one 10-ml tube can sometimes be used to obtain enough serum for several tests.*

PERSONNEL

To have a successful operation, adequate staffing of the laboratory is essential. The technicians hired for the processing of specimens must have a good memory and be able to work quickly and without constant supervision. They also must be able to work in proximity with a large number of other people (in larger institutions) or alone when on rotating shifts, be able to work and communicate with employees, physicians, and laboratory personnel, and possess good human relations skills.

The technician must learn to operate numerous pieces of equipment, such as centrifuges, refrigerated centrifuges, pneumatic tubes (if available), and Vortex mixers. Training is accquired on the job and is conducted by an experienced technician designated as trainer. Usually training takes 6 to 8 weeks to learn the basic functions and duties. Approximately 1 year is required to perform the job proficiently in all respects. The learning process is continuous because test changes are being made frequently.

PHYSICAL LAYOUT AND EQUIPMENT

Assumptions involved in the design configuration for a central processing facility (Fig. 8-1) include

- scope of activity to be undertaken by the laboratory;
- degree of personnel and equipment specialization to be incorporated into the design concept; and
- type of accessioning system or systems to be used by the laboratories.

Fig. 8-1. *General view of the Central Processing Laboratory at the Mayo Clinic.*

The design and layout characteristics necessarily depend on the type and quantity of specimens processed. The facility should encompass the following stations (Fig. 8-2).

Receiving and sorting
Accessioning
Centrifuging
Aliquoting
Specimen storage
Delivering station
Emergency area
Inquiry station

CENTRIFUGES

Two types of centrifuges should be available: regular centrifuges for specimens that may be processed at room temperature, and refrigerated centrifuges for specimens that should be processed at lower temperatures. The number of centrifuges needed depends on the volume of specimens to be processed in the laboratory. In our institution, where we process more than 12,000 specimens a day, we have 22 regular bench-top model centrifuges and one refrigerated centrifuge.

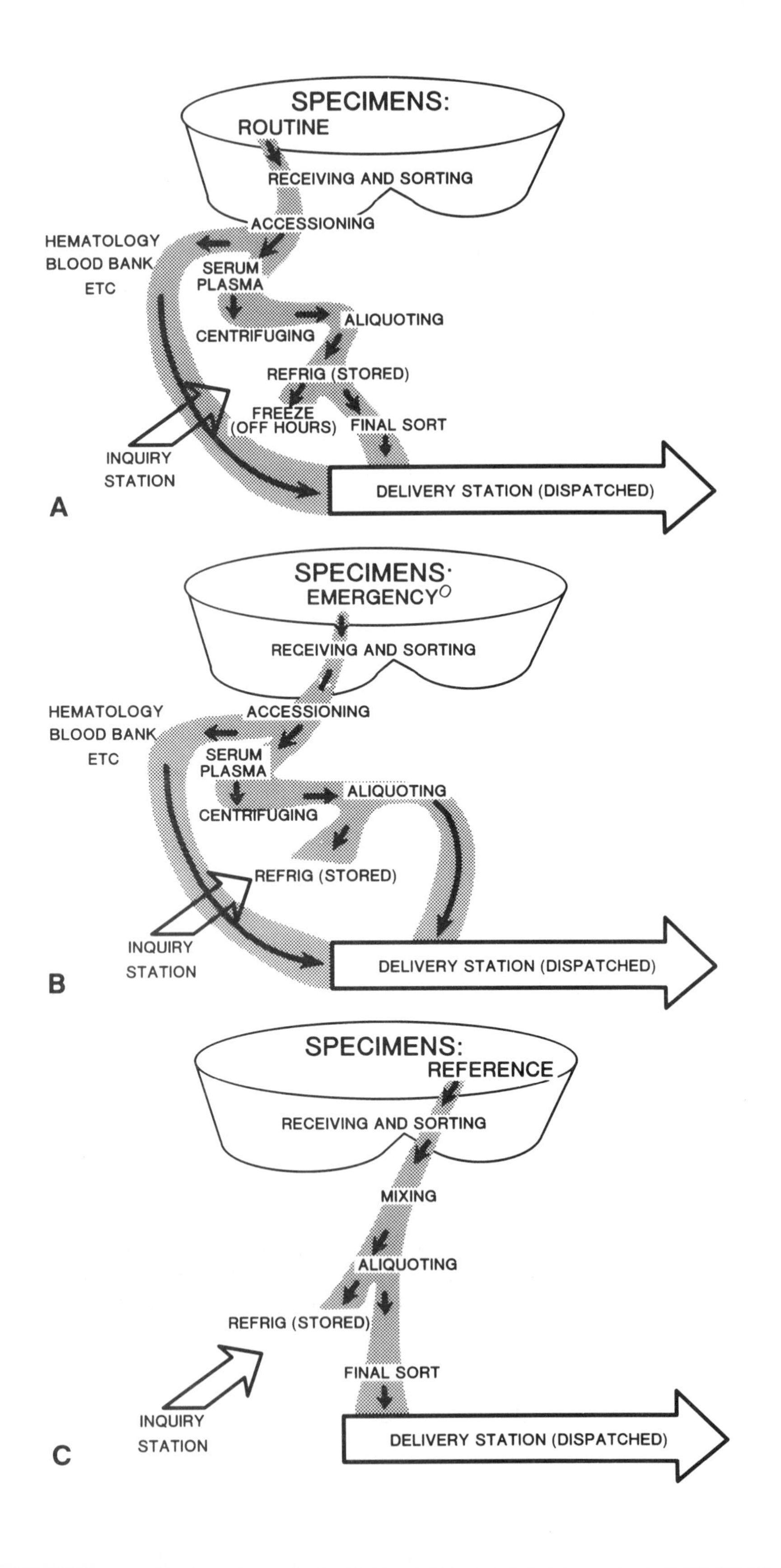
SPECIMENS:
ROUTINE
RECEIVING AND SORTING
ACCESSIONING
HEMATOLOGY
BLOOD BANK
ETC
SERUM
PLASMA
CENTRIFUGING
ALIQUOTING
REFRIG (STORED)
FREEZE
(OFF HOURS)
FINAL SORT
INQUIRY
STATION
DELIVERY STATION (DISPATCHED)
A
SPECIMENS:
EMERGENCY
RECEIVING AND SORTING
ACCESSIONING
HEMATOLOGY
BLOOD BANK
ETC
SERUM
PLASMA
CENTRIFUGING
ALIQUOTING
REFRIG (STORED)
INQUIRY
STATION
DELIVERY STATION (DISPATCHED)
B
SPECIMENS:
REFERENCE
RECEIVING AND SORTING
MIXING
ALIQUOTING
REFRIG (STORED)
FINAL SORT
INQUIRY
STATION
DELIVERY STATION (DISPATCHED)
C

COMPUTER TERMINALS

The computer system used in the medical laboratory should be able to provide many functions that will help the technician to perform the varied daily operations. The functions can be performed at a cathode ray tube (CRT) terminal. Some of the functions that the system should be able to do include system inquiry, order entry, test inquiry, deleting, sample status, sample label, test library, and patient identification.

The computer equipment in a central processing laboratory would include CRTs and printers. One CRT and printer would be used for emergency specimens; another CRT and printer would be used for routine specimens. Two CRTs would be used for chemistry group specimens (one to generate the worksheet and the other to display the worksheet for reference in the aliquoting area), and a CRT in the inquiry area would help to solve problems that arise and retrieve information on stored serum.

BEAD DISPENSERS

The bead dispensers (Fig. 8-3) are not essential equipment in a central processing laboratory. If available, they help to elminate the rimming of blood tubes and help to deliver a higher yield of serum, free of fibrin debris. The bead dispensers are mechanical devices that dispense 15 to 25 small polystyrene beads, sieve size 16, into the blood tube. The technician activates the dispenser when the tube is in place by tapping a foot pedal.

REFRIGERATORS AND FREEZERS

Refrigerators are used to store serum or plasma at 4°C and to store the clot tubes. Many tests are stable for 48 hours when the serum is left in the tube over the red blood cell clot. The tests that cannot be run if the serum is left over the clot are glucose, potassium, lactic dehydrogenase (LDH), iron, alanine aminotransferase (ALT), and aspartate aminotransferase (AST).[2]

Freezers are used to store serum or plasma specimens that arrive during off-hours and need to be frozen until they can be tested. Any whole blood should be centrifuged and the serum or plasma frozen.

DISTRIBUTION SYSTEMS

It is ideal to have a good mechanical system for the delivery of specimens. Examples of these systems are conveyor belts, lifts, pneumatic tubes, and Rally Post. A good back-up system should be available, such as messengers.

At our institution we have the following delivery systems.

- Pneumatic tube—two systems are used, the point-to-point system and the system that connects to all points of the Mayo Clinic complex. All emergency specimens arrive through pneumatic tube from their point of origin.

◀ **Fig. 8-2.** *Flow diagram of specimen flow for* (A) *routine specimens,* (B) *emergency specimens, and* (C) *reference specimens.*

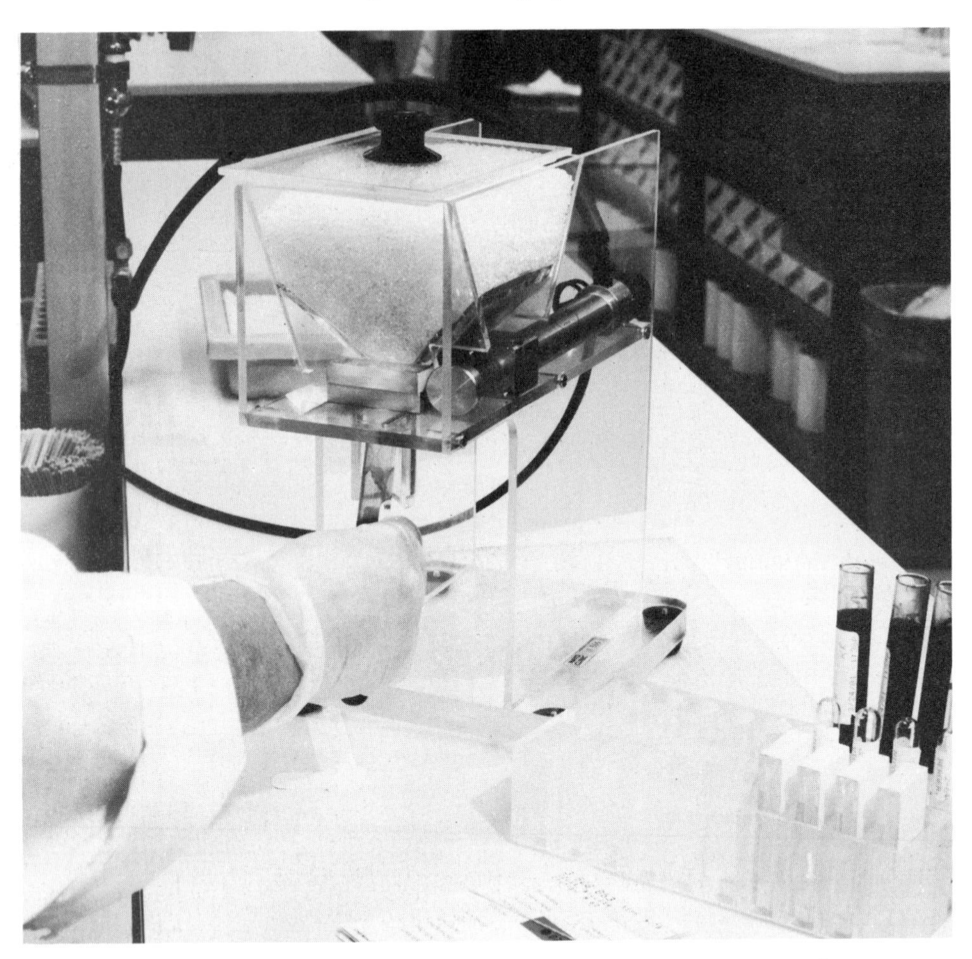

Fig. 8-3. *Bead dispenser designed by the Mayo Clinic Engineering Department to dispose 15 to 25 beads at a time.*

- Conveyor belts—point-to-point conveyor belts (Fig. 8-4) connect our central processing area with the outpatient venipuncture area. Two racks can be placed on each belt, and each rack can hold blood from three patients.
- Rally Post—used to deliver specimens to the different laboratories throughout the building. Fifty battery-operated cars follow electronically switched tracks throughout the building. We can send cars to 11 laboratory stations on four different floors.

PROCESSING PROCEDURES

An up-to-date procedure manual should be available to all central processing staff for processing blood specimens. The manual should contain every test

Fig. 8-4. *Conveyor belt is used to bring samples from the outpatient's venipuncture area to the Central Processing Laboratory. Two racks can be placed on the conveyor that hold specimens for six patients.*

performed in the department, listed alphabetically. If a test is known by more than one name, a cross-reference index should be kept. The procedure manual should be very complete and kept up-to-date because it will be used 24 hours a day, 7 days a week.

TEST STAFF ______________ SUPERVISOR ______________

NAME LDH Isoenzyme ROOM #342H RALLY POST 230 Phone EXT 3213

LAB HOURS 8 am–5 pm Monday–Friday

ANTICOAGULANT None TUBE SIZE 10 ml ALIQUOT AMNT 2 ml

MINIMUN AMOUNT ______________

DURING REGULAR HOURS Sort into a chemistry block, centrifuge, aliquot, and send to the laboratory.

DURING OFF HOURS Centrifuge, aliquot, and store in the refrigerator until next regular working day.

SPECIAL INSTRUCTIONS

CENTRAL PROCESSING LAB

For each test, a sheet should be prepared that contains detailed processing and handling information—for example, the test name, the laboratory performing the test, the laboratory's hours, the anticoagulant, tube size, aliquot amount, minimum amount acceptable, rejection criteria, and special factors involved.

A well-documented quality-control procedures manual should be available and have all the quality-control assurance procedures used throughout the laboratory, the safety procedure, equipment function and verification procedures, maintenance, and cleaning procedures. Records of quality-control activities should be reviewed regularly and filed in the manual.

REVIEWING AND SORTING OF SPECIMENS

The process of identifying (Fig. 8-5) and matching specimens is sophisticated today. In the past, day numbers and colored tapes were the only means of identifying and matching. Today, computer labels provide a computer-assigned number, clinic or hospital number, and the patient's full name. Computer labels have helped to reduce the number of errors that used to occur with the day-number system, such as people forgetting to turn the knob that indicated the day number or, in aliquoting, duplication of the colored tape numbers or someone forgetting to place the colored tape numbers on the tubes.

Matching and sorting blood can be accomplished by segregating specimens by laboratory, by function of operation, or by type of specimens. Before dispersing specimens, the persons assigned to the sorting and receiving bench must be trained in the following functions.

1. To check the patient's full name, registration number, and specimen computer number on the blood tubes and the test registration cards to be sure that they match.
2. To check that the right type of specimen has been sent for the test ordered and that the right preservative or anticoagulant has been used.
3. To recognize which test samples must be iced and which must be centrifuged in the refrigerated machine.

4. To know the procedure that must be followed if the specimen is not labeled correctly or if the wrong type of specimen is taken.
5. To be familiar with all types of vacuum tubes and microtubes used to collect blood, plus all containers for special fluids.

Specimens not needing any centrifugation or processing, such as hematology specimens, can be sent immediately to the laboratory after matching. All blood specimens are received in this area. Outpatient specimens are carried by hand or conveyor. Hospital specimens may arrive by pneumatic tube or special system for the emergency work and by delivery service for the routine work. Specimens are sorted into racks depending on the degree of operation and laboratory requirements. Specimens are routed according to urgency.

The sorter must have certain knowledge of each test: whether anticoagulants must be used, whether the specimens require special processing such as refrigerated centrifugation, and which laboratories receive what specimens. Before putting the blood into the racks, the sorter must check the name and clinic number of the patient to ascertain that the tubes and labels all match the patient and to be sure that the additional labels are sent along with the blood for use by the laboratories.

PRECENTRIFUGATION

Precentrifugation is the time period after a specimen has been collected and before it is centrifuged.

Fig. 8-5. *Identification and labeling of specimens before and now. Two samples on the left show the old system of using the day number. Three samples on the right show the current system of using labels with computer number, hospital clinic number, and the patient's full name.*

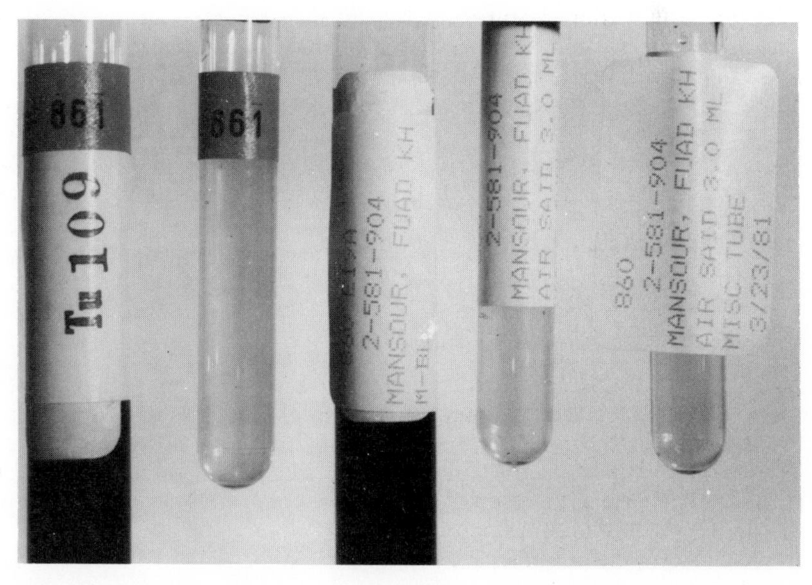

Serum Versus Plasma

Blood, when withdrawn from the body and placed in a clean tube without any preservative, forms clots by converting soluble protein—fibrinogen—to insoluble protein—fibrin. The fibrin forms a spongy network of fibrous material throughout the body, holding the blood corpuscles into a solid mass. Clotting usually takes 10 to 15 minutes, depending on the amount of blood drawn and the type of tube used.[3]

The difference between serum and plasma is that serum is formed by clotted blood, which contains fibrin, and plasma, which contains fibrinogen, is formed from blood that is not allowed to clot because of an anticoagulant. A quick test, the fibrinogen test, can be used to determine whether a specimen is serum or plasma: With serum, there is no reaction. Plasma anticoagulated specimens can be centrifuged as soon as they are collected. Serum specimens, blood with no anticoagulants, must be allowed to clot for at least 30 minutes at room temperature. If the collection tubes contain a clotting activator, such as glass or silicon particles, the clotting time can be shortened to 15 minutes.[4–7] If thrombin is used, clotting time can be reduced to 5 minutes.[8]

If the blood is drawn and allowed to stand in a refrigerator or on ice, it requires a longer time to clot. Allowing the clot to retract for a longer time minimizes hemolysis and increases the yield of serum. Glycolysis, however, can take place with a shift of substances from the cells to the serum. Potassium is one of the substances that can be effected by glycolysis. Drawing the blood into an evacuated tube minimizes hemolysis and prevents the clot from attaching to the walls of the tube.[9]

Removal of Stoppers

After the blood has clotted, the stopper should be removed with a special device available commercially. The technician must wear gloves to perform this task for safety reasons, to keep the hands clean, and to protect the hands in case of tube breakage. In removing the stopper, the technician should hold the tube away from the body to avoid splashing. Moreover, the stopper should be removed carefully because disturbing the clot can cause hemolysis.

Rimming

The rimming of blood with a wooden applicator stick helps to loosen the clot and to separate it from the walls of the tube, although excessive rimming and agitation of the specimen can damage the clot and cause hemolysis. Rimming also tends to be messy. Recently, the use of tube separators or beads has reduced the use of rimming sticks.[4, 9]

Polystyrene Beads

Polystyrene beads may be used after the blood clots to keep the serum free from fibrin debris and to help keep the cell packed. If large volumes of specimens are to be processed, a dispenser should be designed. Our engineering department found no difficulty in designing a suitable dispenser. The design was similar to one that they had developed for the pediatric department for dispensing M & M candies to children.

We recommend that technicians use protective gloves when dispensing the beads into tubes. The dispensers should be kept clean, and we do not recommend that the blood tubes be allowed to touch the mouth of the dispenser. We do not use the beads for emergency specimens because they need to be processed right away, and by beading them an unwanted delay occurs.

Separator Types

Over the years, many different types of separators have been used to process specimens. Calam details the products used to assist in the processing of specimens.[4] The first group of gels used to separate sera from cells was found to be inefficient or containing contaminants that affected the test results. Recently, however, a new generation of serum separator tubes has appeared on the market, including the SST of Becton–Dickinson,[5] AutoSep of Venoject,[7] and Corvac of Monoject.[6] Before these new products are used, however, they should be evaluated thoroughly to be sure that no contamination exists or that there is no effect on the specimen that could cause tests error. A detailed evaluation should include a change in test results, clotting time, volume of serum, and quality-control factors such as sterility, vacuum, breakage, stopper removal, ability to pour off specimen versus pipetting, temperature effect, and cost.

CENTRIFUGATION

Bloods are placed into the centrifuge in such a way that all specimens are balanced. The same sized tubes should be placed exactly opposite each other. If an odd number of specimens is to be processed, water tubes should be used as a balance. One tube alone should never be centrifuged. Blood tubes are centrifuged at $1100 \times g$ for 10 minutes using a Sorvall GLC II centrifuge. The procedure manual about the specifications for centrifugation provided by the manufacturer with each machine should always be read.

Centrifuging bloods with the stopper on is widely practiced because it reduces the evaporation caused by air forces.[4] If the stopper is left on during centrifugation, care should be taken when removing it after spinning. Any agitation could cause the packed cells to disrupt, thus necessitating recentrifugation of the specimens, causing loss of time. The stopper must be kept on during centrifugation for the following tests.[4]

- Anaerobic specimens (blood gases, *p*H, carbon dioxide, ionized calcium, acid phosphatase)
- Volatile analyte specimens (ethyl alcohol, ketone bodies)
- Readily oxidized specimens (vitamin A, carotene)
- Microbiology specimens (antibiotics)

Recently, centrifuge manufacturers have introduced a special sealing cover that can be placed over the buckets. When this device is used, there is no need to keep most tubes stoppered during centrifugation, except for the tests listed above. This type of covered bucket minimizes aerosol formation should a tube shatter during centrifugation. Also, it is easier to clean the centrifuge should breakage occur because debris would be isolated in the bucket, not strewn throughout the whole centrifuge.

Certain stoppers that contain the compound tris butoxyethyl phosphate have been found to interfere in tricyclic drug analysis. This compound, used as a lubricant, dissolves in the heparinized plasma and causes lower test results for such tests as Elavil, Aventyl, Tofranil, Norpramin, and Sinequan. This type of stopper should *not* be used. Recently, some manufacturers have stopped making the tris butoxyethyl phosphate stopper. Some heparinized tubes, however, contain the lubricant; they too should not be used.

ALIQUOTING

Preparation of Aliquot Tubes and Vials

While the blood tubes are centrifuging, the technician should set up the aliquot tubes (12 mm × 75 mm) for each patient's serum or plasma. If computer labels are used for labeling the aliquot tubes, an extra computer label can be placed on the edge of the rack next to the tubes for a quality-control check. The computer labels used for aliquoting should have the patient's full name, computer-assigned number, clinic or hospital number, name of the test, and volume required to run the test. Aliquot tubes can be glass or plastic; however, because glass tubes can break, polystyrene tubes may be better for safety reasons. It is

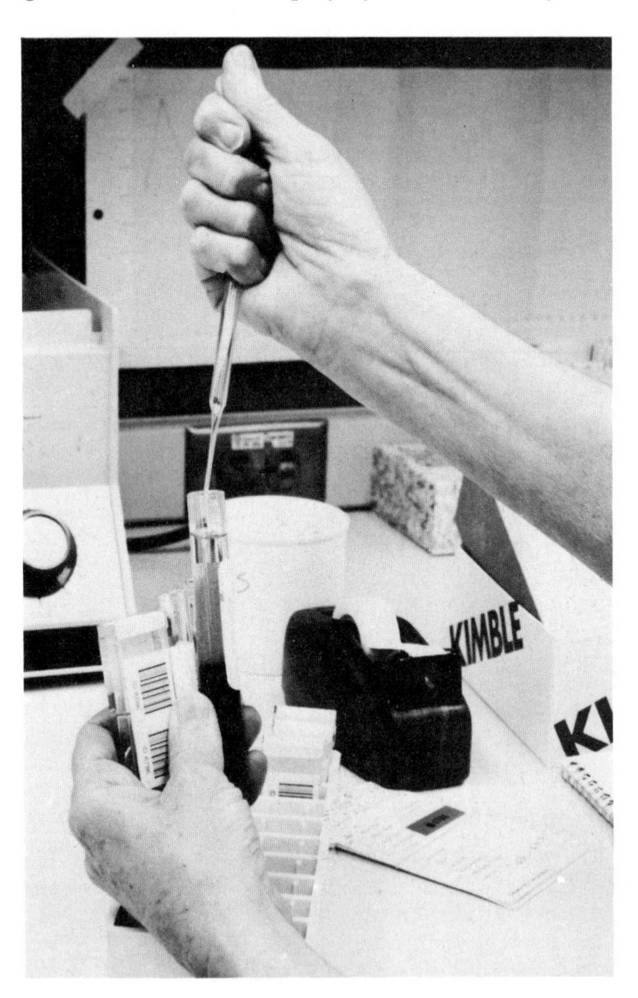

Fig. 8-6. *Pipetting of samples using disposable pipette and rubber bulb.*

important to test the tubes used for aliquots to make sure that they contain no trace metals or leachates.

Pouring Versus Pipetting

When the blood has been centrifuged, serum is aliquoted using disposable metal-free pipettes (Fig. 8-6). Pouring is a bad technique to transfer the serum or plasma to the aliquot tube from the blood tube because the packed cells are easy to agitate and some cells get in with the serum or plasma. There may also be contamination when samples are poured, especially for lipid tests. Many technicians use hand creams, which are full of oily substances. As they pour the serum or plasma, some of the cream may wash out easily with the substance and cause elevated test results. Other contamination may occur that affects electrolyte, sodium, and potassium results by pouring with sweaty hands.

QUALITY CONTROL

Technicians assigned to aliquot specimens (Fig. 8-7) have a special form for quality control (Fig. 8-8) with a place for their name, the date, and the area or

Fig. 8-7. *Aliquoting of specimens. The rack is used for the sample of aliquots and the extra labels at the edge are used for a quality control check.*

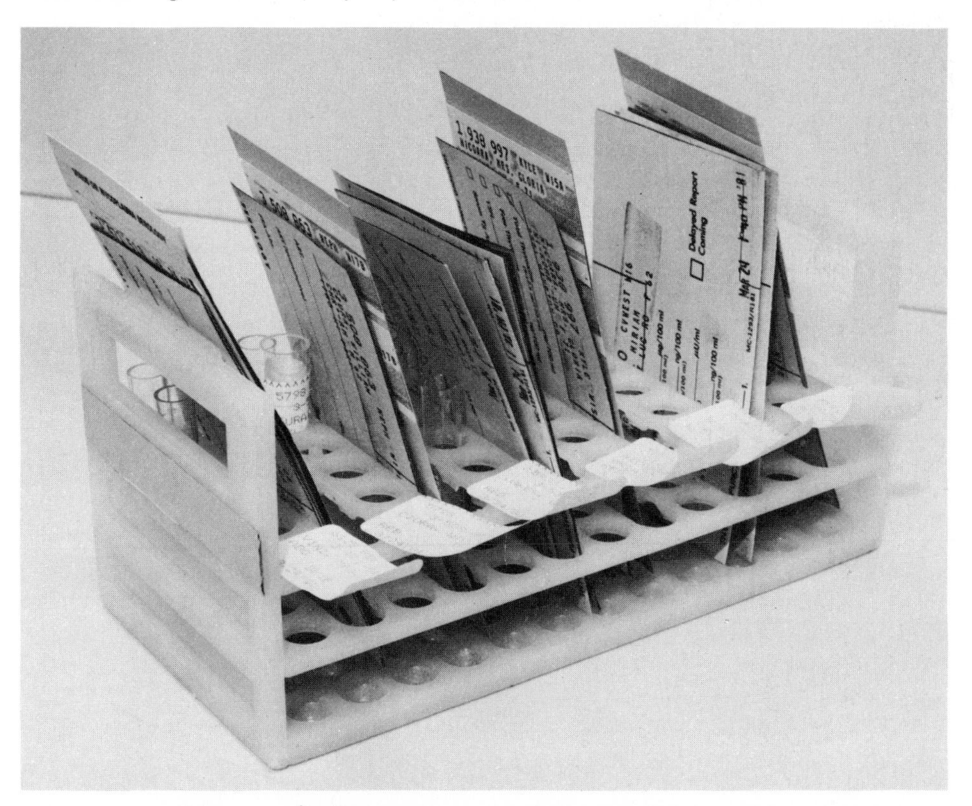

1949 ENZYMES
2-598-437
SONNENBERG, MRS. CHE 5/45F
CKI CK LDI LD
E19A 4.0 ML
7/ 6/82 3:40P

1952 ENZYMES
2-581-904
MANSOUR, FUAD KHAIR 2/41M
CKI CK LDI LD
W4B 4.0 ML
7/ 6/82 3:40P

1953 ENZYMES
2-405-166
GAMM, MRS. MARJORIE 11/24F
CKI CK LDI LD
E19A 4.0 ML
7/ 6/82 3:41P

1956 ENZYMES
2-005-560
KOEBKE, MRS. JEAN 9/37F
CKI CK LDI LD
W4B 4.0 ML
7/ 6/82 3:41P

1957 ENZYMES
1-143-913
MCFARLAND, MRS. ESTH 9/23F
CKI CK LDI LD
W18 4.0 ML
7/ 6/82 3:41P

1949 E19A
2-598-437
SONNENBERG, MRS. CHE 5/45F
CKI CK LDI LD
7/ 6/82 3:40P

1952 W4B
2-581-904
MANSOUR, FUAD KHAIR 2/41M
CKI CK LDI LD
7/ 6/82 3:40P

1953 E19A
2-405-166
GAMM, MRS. MARJORIE 11/24F
CKI CK LDI LD
7/ 6/82 3:41P

1956 W4B
2-005-560
KOEBKE, MRS. JEAN 9/37F
CKI CK LDI LD
7/ 6/82 3:41P

Fig. 8.8. *Example of sheet used for quality control by technicians as they aliquot each patient.*

bench where they are aliquoting. As they pipette and aliquot the serum, technicians will take the extra label placed on the edge of the rack, check to make sure that it matches the vial labels, then pipette, aliquot, and place the label on the quality-control sheet. This technique ensures that the technician is pipetting the right patient's blood, which reduces the error of mismatching. The sheets from each technician are saved for 1 week. In case of problems or errors, one can identify the person responsible by checking the quality-control sheets.

For the chemistry group, a special rack (Fig. 8-9) helps to keep the clotted blood, stored serum tube, and instrument vial in one rack. Each rack holds blood specimens for ten patients. The technician matches the blood with the stored serum tube and the instrument vial before aliquoting. After pipetting the serum of each patient, the technician initials a special quality-control book that indicates the ten specimens processed. Another technician double-checks the finished rack to ensure that all specimens match. The second technician also initials the quality-control book next to the first person.

The rack used for this procedure is made in such a way that the instrument serum vials can be separated from the clot tubes and the stored serum tube. The instrument tubes are sent to the laboratory for testing. The clot and stored serum are held in the refrigerator for 1 week. If there are any problems or errors, one can check the clot tube and the stored serum tube to see if they match—another quality-control check in the operation.

EMERGENCY SPECIMENS

Special priority should be assigned to emergency specimens. A laboratory must have a well-defined protocol to follow in the handling and processing of emergency specimens. These specimens should be handled in a special area with enough help to keep the specimen flowing without any delay. The specimen would arrive by special messenger delivery or by pneumatic tube. Enough centrifuges should be available to keep centrifuging the specimens.

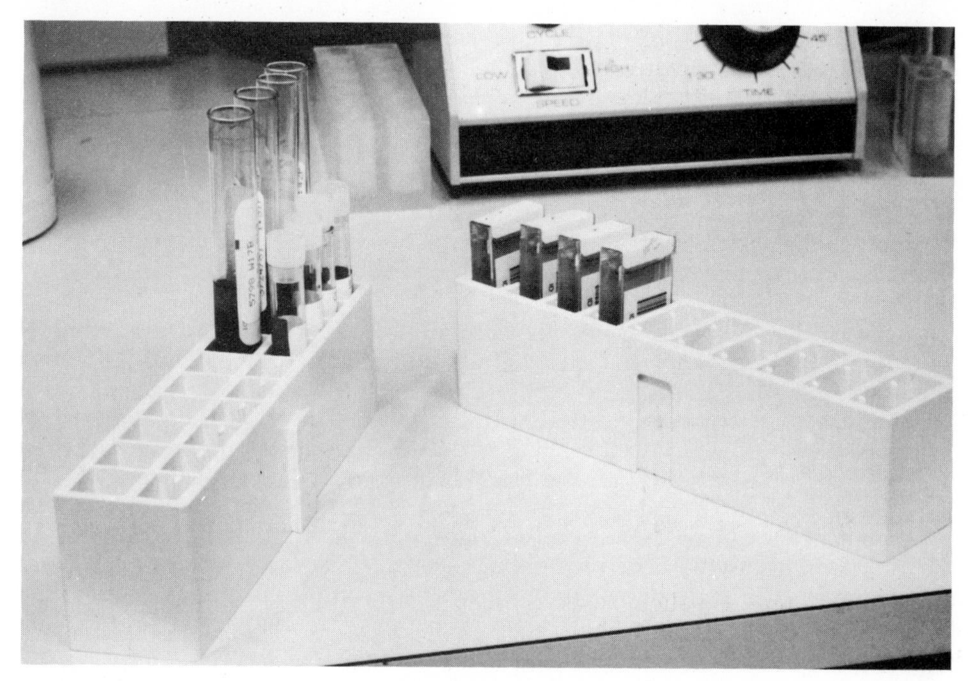

Fig. 8-9. *Special racks for chemistry group specimens. Shown is the rack when it is connected and when it is separated. It is used as a third quality-control check if there is a mismatch.*

Two sets of priorities should be available: first-priority specimens, in which results must be completed within 1 hour of the time of drawing to the time of reporting the results; and second-priority specimens, in which the results must be reported within 3 hours of the time of collection.

The following functions are performed.

- Sorting and handling should be performed without delay. The specimen must be matched with the patient's hospital and clinic numbers.
- After sorting and matching, the specimens are processed according to tests. Hematology, urinalysis, and blood-bank specimens are sent immediately to the appropriate laboratories. The chemistry specimens are centrifuged for 5 minutes. During centrifugation, the aliquot tubes are prepared. Serum or plasma is aliquoted for each test ordered. Special red racks are used to identify medically urgent specimens. Also, labels on the request slips are color coded to identify the emergency specimen.

EMERGENCY-ROOM SPECIMENS

Specimens from the emergency room take priority over all other work and should be processed separately according to the following procedures.

1. Initial handling and sorting of the specimens: Match blood samples with the test forms, checking full name, clinic or hospital number, and computer (sample) number. Check that the right type of specimen is collected for each test.
2. Centrifugation: Centrifuge the specimen right away and set the timer for 5 minutes.
3. Aliquoting: As the specimens are being centrifuged, prepare the aliquot serum tubes. After centrifugation, aliquot the specimens and send them to the appropriate laboratory as quickly as possible in specially marked racks.

A specially designed system is needed for specimens from unidentified patients in the emergency room to identify and match these patients correctly. An alpha numeric system could be used to label these bloods: It contains four-digit numbers and comes in different colors. The numbers will be attached to the wrist of the patient. As the blood is drawn, the alpha numeric number is used to match the patient with the blood tests.

It is important to find out how long it takes to process emergency-room specimens. This can be done by time-stamping when the specimens arrive at the laboratory and when the processing is complete and the specimens leave.

At our institution the time study is done on a continuous basis to determine the processing time of emergency room specimens. Timing of emergency-room specimens for the month of May 1981 is as follows.

Shift: 12:00 a.m.—7:00 a.m. Avg. # of Patients 6
Avg. Time 8.38 Avg. # of Tests 51 Total Tests 1575
Avg. Tests/Patient 9 St. Mary's 127 Methodist 67
Avg./Day 4 2

Shift: 7:00 a.m.—3:30 p.m. Avg. # of Patients 16
Avg. Time 11.06 Avg. # of Tests 139.5 Total Tests 4326
Avg. Tests/Patient 9 St. Mary's 305 Methodist 197
Avg./Day 9 6

Shift: 3:30 p.m.—12:00 a.m. Avg. # of Patients 23
Avg. Time 11.32 Avg. # of Tests 187.8 Total Tests 5823
Avg. Tests/Patient 9 St. Mary's 458 Methodist 249
Avg./Day 14 8

Total Patients 1403 Total Tests 11724 Highest # of Tests 23
St. Mary's Total 890 Methodist Total 513 Avg. Time/Day 10.52
Avg. # of Tests/Day 378.2 Avg. # of Patients/Day 44

FROZEN SPECIMENS

Certain tests require that the specimen be stored frozen during off hours to maintain the test's stability. Off hours refers to the hours when the laboratory that performs the test is closed. Some laboratories require that the specimens be delivered frozen on Dry Ice the next working day, and some request that the specimen be thawed before delivery. If a specimen is stored frozen, it must be mixed thoroughly on a Vortex mixer when thawed.

The following tests need specimen freezing off hours.

Angiotensin-converting enzyme (ACE)
Aldosterone
Adrenocorticotrophic hormone (ACTH)
Carotene
Catecholamine
Carcinoembryonic antigen (CEA)
Complement (Total,C3,C4)
CT assay (calcitonin—plasma in a PTH vial)
Digoxin
Estrogen
Ferritin
Follicle stimulating hormone (FSH)
Haptoglobin
Human growth hormone (HGH)
Immunoglobulins
Insulin
Insulin antibodies
Inborn errors—urine
LH (Lutenizing Hormone)
Methotrexate (MTX)
Prolactin
Phytanic acid
PTH (parathyroid hormone)
Testosterone
Renin
Vitamin A
Vitamin B_{12} and Folate

Seldom does one need to freeze the whole blood tube in the freezer. If the whole blood is frozen, hemolysis could occur and the glass tube could break. It is always important to centrifuge the blood and separate the serum or plasma and place it in polypropylene vials for freezing.

CRITERIA FOR ACCEPTING OR REJECTING SPECIMENS

Each laboratory must have its own well-defined criteria for accepting and rejecting specimens. If these criteria are not followed, inaccurate test results may occur.

The following are the most common conditions that can cause specimens to be rejected.

- Specimen misidentification or unidentified specimen.
- Inadequate volume of blood collected into a tube with an anticoagulant.
- Specimen collected in the wrong collection tube.

- Temperature. For certain blood tests, the specimen should be cold always. The following tests should stay on ice from the time of collection until analyzed: prothrombin, APTT, acetone, carbon dioxide (acids/base), ammonia, free fatty acids, ACTH, amrinone, B-lipotropin, catecholamines, fibrinogen, glucagon, isovaleric acid, uroporphyrinogen 1 synthase, dopamine, renin, gastrin, and calcitonin. Whole blood specimens that are refrigerated or held on ice are unacceptable for potassium determinations.[10]
- Hemolysis. When hemolysis occurs the specimen is a reddish color. The red blood cells are lysed, and the hemoglobin is released into the remaining portion of the blood. Hemolysis can be caused by venipuncture, dirty tubes, tubes that are not dry, freezing, and pipetting the blood too vigorously into the aliquot tube.[11] Hemolysis can affect the test results of potassium, phosphorus, bilirubin, lactic dehydrogenase, and acid phosphatase.
- Repeated freezing and thawing of the specimen.
- Insufficient amount of the specimen.
- Lipemic specimens.
- Specimens from patients who are not fasting for those tests where fasting is necessary.

INCIDENT REPORTS

To help solve problems that arise in specimen handling and processing, the use of a special error report can help identify and reduce problems. Some errors that can be identified from the incident reports include

- no label identification on blood tubes;
- mismatched specimen;
- mismatched aliquot;
- misplaced specimen;
- errors in collecting, handling, and processing;
- delayed specimen; and
- delivery problems.

Each laboratory should develop its own incident form to fit specific needs. The form should include the following.

- Patient name and registration number
- Date and time of incident
- Person involved (physician, supervisor, technician)
- Description of incident
- Signature of person filing report
- A recommendation of how to avoid problem in future

Periodically, the reports should be reviewed to be sure that no one type of incident is occurring too often. The results of the reviewals can then be used to help improve the system, resulting in fewer errors.

One should test the system occasionally to determine how long it takes to process the specimens. Statistics in our laboratory for 1981 show the following results (Table 8-1).

Table 8-1. *Specimen Processing Time, Central Processing (Time in Minutes), 1981*

LABORATORY		JAN	FEB	MAR	APR	MAY	JUN	JUL	AUG	SEP	OCT	NOV	DEC
Lipid	*Avg.*	50.0	51.9	55.5	53.6	51.5	54.6	58.1	50.4	60.7	57.1	60.6	61.2
	SD	13.0	18.9	15.2	21.0	13.9	10.5	9.1	10.9	12.8	12.4	9.4	10.7
Vickers	*Avg.*	43.6	51.9	49.5	44.4	38.9	49.4	52.2	51.7	56.0	54.6	55.2	55.3
	SD	17.4	15.3	18.0	20.3	18.5	8.3	13.5	9.0	10.6	11.7	8.7	10.6
Drugs	*Avg.*	50.6	47.7	53.9	44.9	53.3	48.2	60.9	57.8	64.2	59.9	57.2	54.7
	SD	14.6	21.0	19.7	17.6	17.3	12.3	11.1	8.1	12.5	15.6	15.9	11.1
Thyroxine	*Avg.*	47.2	58.5	56.0	58.1	53.4	53.5	61.8	57.4	61.8	57.0	60.9	60.4
	SD	20.0	20.3	17.5	16.3	12.9	13.3	10.7	9.2	6.8	8.7	8.4	8.9

Centralization of specimen processing is a unique system. Because of the large volume of specimens sent to one location, specimens must be processed rapidly and without delay. A realistic objective should be to complete the processing of specimens within 1 hour from the time it enters the processing area to the time it is ready for delivery to the final testing laboratory.

REFERENCES

1. Procedure Guide for Mayo Clinic Physicians, p 4–1. Rochester, Minnesota, Mayo Clinic, 1980
2. National Committee for Clinical Laboratory Standards (NCCLS): Standard for Handling and Processing of Blood Specimens, H18-P. Villanova, NCCLS, 1982
3. MacFate RP: Introduction to the Clinical Laboratory, 3rd ed, pp 9–10. Chicago, Year Book Medical Publishers, 1972
4. Calam RR: Reviewing the importance of specimen collection. Am Med Technol 39:297–302, 1977
5. Serum Separator Tube™. Performance Data (on file). Rutherford, New Jersey, Becton–Dickinson Co
6. CorVac® Integrated Serum Separator Tube. Performance Data (on file). St. Louis, Monoject, Division of Sherwood Medical
7. Auto Sep™. Performance Data (on file). Elkton, Maryland, Terumo Medical Corp
8. Steindel SJ: Evaluation of a thrombin containing blood collection tube. Clin Chem 26:173–174, 1980
9. Tietz NW: Fundamentals of Clinical Chemistry, 2nd ed, pp 47–52. Philadelphia, 1976
10. Oliver TK, Young GA, Bates GD, Adams JS: Factitial hyperkalemia due to icing before analysis. Pediatrics 38:900–902, 1966
11. Linne J, Ringsrud KM: Basic Laboratory Techniques for the Medical Laboratory Technician, pp 68–71. New York, McGraw–Hill, 1970

Part IV

GENERAL SPECIMEN CONSIDERATIONS

9

How Laboratory Tests Are Used In Clinical Medicine

Michael B. O'Sullivan

During the past 30 years the role of the clinical laboratory in the practice of medicine has changed dramatically. Before the 1950s, clinical laboratories were quite primitive by today's standards. Reliable laboratory tests were relatively few and were performed tediously by the manual efforts of technologists. Because physicians today rely more and more on laboratory tests, their volume has grown enormously. This growth has been facilitated by modern science and technology. Manual tests have given way to automation so that highly sophisticated analytical instruments and computers are now the tools of the modern clinical laboratory. In addition, through medical research, a whole new and vast array of tests have been put at the disposal of the physician to help in the diagnosis, treatment, and monitoring of disease. By selecting the proper tests, a physician can often pinpoint quite precisely what is going wrong in one of the body's many and complicated biological systems. For instance, chemical determination of blood urea nitrogen or creatinine may provide the first clue of early kidney failure, while measurement of the numbers of red cells or white cells in the blood may indicate that a patient has anemia, infection, or leukemia.

Physicians order laboratory tests for numerous reasons (Table 9-1).

Table 9-1. *Reasons for Ordering Laboratory Tests*

1. To diagnose disease in sick patients
2. To screen for disease in seemingly healthy persons
3. To assess the extent of disease or degree of body damage
4. To follow the course of disease
5. To assess the effectiveness of treatment
6. To monitor drug treatment levels

DIAGNOSIS

In reaching a diagnosis, physicians know that each test they select is designed to answer specific questions. For instance, if a physician determines that a patient's tiredness and lack of energy are due to anemia, that is, a low blood count with insufficient hemoglobin, then he or she will attempt by further laboratory testing to determine the exact cause of the condition. Does the patient lack enough iron or sufficient vitamins to manufacture healthy red cells, or is the patient bleeding continuously but in such small amounts that it is not readily apparent? There are tests to measure the levels of serum iron, vitamin B_{12}, and folic acid, and a number of laboratory determinations can be done to determine if the patient is losing small quantities of blood from the stomach or bowel.

At the annual physical examination, the physician may order a series of laboratory tests to ascertain what is happening in the body where the stethoscope, x-ray, and physical examination cannot probe. Disease often begins slowly and imperceptibly without any obvious symptoms or signs, but changes in the blood or in body cells may warn of approaching disease in time for preventive action. For example, a high serum cholesterol level may tell the physician that a patient is at increased risk for developing a heart attack. A fasting blood glucose determination may reveal the first sign of early diabetes.

When a sick patient is admitted to the hospital, laboratory tests not only will help in identifying the cause of illness but also may help in determining the extent of disease. For instance, in a patient suffering from hepatitis, enzymes from the injured liver tissue may leak into the blood, a fact that is helpful in diagnosis and also in determining the extent of damage. Higher enzyme levels generally signify more extensive liver damage. Such information is obviously of great importance to the physician.

The physician not only may request laboratory tests at the patient's initial admission to the hospital to determine the cause and extent of the patient's illness, but he or she may order follow-up tests to ascertain whether three is healing or progression of disease. In the previous example of the patient with hepatitis, falling enzyme levels would be a good omen, whereas remaining high or increasing levels may indicate further progression of liver damage. In chronic diseases, for which long-term treatment may have to be maintained, the physician may need to monitor the patient's progress. In rheumatoid arthritis, patients are moderately anemic, but the degree of anemia usually correlates with the activity of the disease. Similarly, the rate at which the red blood cells settle under the influence of gravity (the erythrocyte sedimentation rate) is generally a good indicator of inflammatory activity in these chronic disease states: The higher the sedimentation rate, the more active the inflammation.

TREATMENT

Effective disease therapy today relies on results of laboratory tests to select the best treatment and then to assess its effectiveness. If a patient has an infection, a culture is done to isolate the organism and identify it with certainty. Then, in the microbiology laboratory, an array of antibiotics are tested for their effectiveness in inhibiting the growth or in killing the organism isolated so that the best antibiotic can be selected to treat the patient's infection. The doctor will want to know how effective the antibiotic is in combating the infection in the patient's system. Subsequent cultures, if negative, will prove that the most appropriate antibiotic therapy indeed was selected.

By monitoring ongoing therapy the physician can fine tune the dose of medication. In diabetic patients, for instance, testing of the urine for glucose or determining blood glucose levels is important in adjusting the dose of insulin. If the patient continues to spill glucose in the urine or if the blood glucose level remains high, the dose of insulin should be increased. Similarly, patients who are given anticoagulants to prevent the blood from clotting are monitored regularly by determination of the prothrombin time. If the prothrombin time is prolonged, it indicates that the therapy is effective and that the blood is less likely to clot in the circulation. If the prothrombin time is too prolonged, however, there is a risk of bleeding. This test is used to monitor carefully therefore the level of the anticoagulants in the blood so that a target level is reached, protecting the patient from the risk of thrombosis and at the same time avoiding too high a dose and the danger of bleeding.

More recently, clinical laboratories have developed the capabilities of measuring actual drug levels in the blood. Although drugs given in therapy have a beneficial effect, most can be toxic, or poison the system, if they accumulate and reach concentrations that are too high in the blood or tissues. Therefore, the ability to measure drug levels in the blood, particularly when toxicity might be life threatening, is of great benefit.

INTERPRETATION

Today a vast array of clinical laboratory resources are available to provide physicians with a scientific basis for making the correct diagnosis of disease and for choosing the best treatment for that disease, even selecting the right dose of medication. Obviously, however, once a physician receives a test result, he or she must be able to determine whether that result is normal or abnormal for a particular patient, that is, whether the result is compatible with health or is more likely to indicate disordered function or disease. With certain clinical laboratory tests that distinction is quite straightforward and uncomplicated. For instance, if blood, urine, or other body fluids properly collected, processed, and cultured grow a pathogenic organism, then that clearly signifies infection. With other laboratory tests, however, the decision may not be as clear-cut. Each human being is an individual, and this quality of individuality applies to the biochemical and cytologic makeup of the body as much as to the more obvious characteristics of physical stature, appearance, and mental and psychological makeup. No two human beings have exactly the same values for all the chemical and hematologic determinations that can be performed on their blood and other body fluids. Whether the results of a laboratory analysis for a patient are normal or abnormal, then, customarily is decided by comparing them with reference values, also called "normal values." Reference values are usually previously determined values from a healthy, screened population. Since age and sex may influence these values, it is helpful with certain tests to report age- and sex-specific normal values. For example, women are known to have lower hemoglobin levels than men throughout most adult life. It is obviously most appropriate, therefore, for a physician to compare the hemoglobin value for a female patient with normal values derived from the female population rather than the entire population.

Many other physiologic and related factors may influence the results of laboratory tests: Examples include body mass and height; geographic location including climate and altitude; activity of reproductive organs including pregnancy, intake of oral contraceptives, and sexual intercourse; food intake and fasting, posture, exercise, stress, diet including caffeine and ethanol intake, hospitalization, travel, tobacco smoking, and drug intake; and normal physiologic rhythms. All of these variable factors must be considered when the physician is trying to determine whether a laboratory test result is normal or abnormal in a given patient. It is obviously a highly complicated decision that is made from the background of extensive knowledge of clinical medicine.

If a laboratory performs a test perfectly on an inadequate specimen, the result may be invalid and lead the physician to a misinterpretation of the result, with possible harm to the patient. For this reason, the highest standards must be practiced in the collection of specimens. Each step, from the physician's initial decision in ordering a test to the collection of the specimen, the performance of a test, and the interpretation of the result is equally important in ensuring that a patient's problem is correctly addressed. By ensuring that blood is collected properly the venipuncturist contributes to the highest standard of medical practice.

SUGGESTED READING

Gornall AG: Basic concepts in laboratory investigation. In Applied Biochemistry of Clinical Disorders. Hagerstown, Harper & Row, 1980

O'Sullivan MB: Laboratory hematology procedures. In Seligson D (ed): CRC Critical Reviews in Clinical Laboratory Science, pp 5–20. Boca Raton, Florida, CRC Press, 1979

Williams RJ: Biochemical Individuality. New York, John Wiley & Sons, 1965

Statland BE, Winkel P: Effects of preanalytical factors on the intraindividual variation of analytes in the blood of healthy subjects: Consideration of preparation of the subject and time of venipuncture. In Seligson D (ed): CRC Critical Reviews in Clinical Laboratory Science, pp 105-144. Boca Raton, Florida, CRC Press, 1977

Young DS: Biological variability. In Brown SS, Mitchell FL, Young DS (eds): Chemical Diagnosis of Disease, pp 1–112. Amsterdam, Elsevier/North Holland, 1980

10

Sources of Variability In Clinical Laboratory Values

James D. Jones

In the past, the major use of clinical laboratories was for confirming diagnoses. Now laboratories are also used in performing large series of screening tests in "healthy" populations to detect subclinical disease and for monitoring patients undergoing treatment or during the course of a disease. The result of these changes in the use of clinical laboratories is the performance of more tests on more people at more frequent intervals. This led to the grouping of laboratory tests for logistical reasons and the production of unsolicited data. Thus, although automation has markedly increased the production of clinical data and availability of additional analytical procedures, it has also created problems.

Aberrant results are frequently attributed to "laboratory error" when in fact they are predictable from other, nonlaboratory factors. A prime objective in laboratory testing is to identify, then eliminate or minimize, the factors that cause unwanted variations and decrease the usefulness of the data produced. In this chapter selected examples of factors that alter laboratory values are documented.

NORMAL VALUES

All laboratory values are interpreted by comparison with values obtained from a reference population. These "normal" or reference values represent 95% limits, that is, 95% of a normal or reference population would have values within the normal value range. The distribution of values within this range is not usually Gaussian, and statistical manipulations are used (log transformations) to determine the high and low limits of the range. The shapes of the distribution curves for normal values for many common tests are given by

Roberts.[1] The normal range can be estimated easily by an X–Y plot of the frequency of a given value versus the concentration or activity.

An institution should, whenever possible, consider all of the factors to be discussed in this chapter when determining any normal ranges. The normal values should be from a control population similar in all respects to the patient population except that they are healthy. The healthy subjects should be treated, instructed, and subjected to venipuncture in the same way as the patients, and the specimens obtained from the patient and the control populations should be handled and analyzed in the same way.

In the absence of a well-controlled healthy population, I have used a similar frequency distribution curve of values obtained from all specimens analyzed including the patient groups under investigation. If calculated on a large unselected population, the distribution values will allow an estimate of a useful reference range; subpopulations are frequently identified by this approach. This method is useful for procedures that require complex patient preparation or difficult specimen collection procedures, for example, on infants.

UNSOLICITED DATA

The introduction of multiple-channel analyzers into clinical laboratories, and with them the screening of selected populations, has created new problems. Although the cost to perform an individual test in a group of tests may be very small, the interpretation of the data and the resulting additional testing caused by an abnormal unsolicited result may be very costly. Most normal ranges are 95% limits, and 5% of the normal population will have an "abnormal" value, that is, one that is outside the normal range. (For unrelated clinical tests, the predicted number of abnormal values can be calculated to equal $[1-0.95^N]$.) This situation most frequently occurs with the introduction of a new test into a battery of tests. A new patient population is being studied, and the data for interpretation are not readily available.

EXTENSION OF TESTING TO NEW GROUPS OF PATIENTS

The increased use of a test leads to automation of the test and application by a larger group of physicians in new populations. If the reference range for the new patient group has not been determined, a physician may be misled. Classic examples include normal ranges that are lower for serum creatinine concentration but higher for creatine kinase and aspartate aminotransferase activities in infants than in adults.

VARIABLES IN LABORATORY TESTING PROCEDURES

Every step in a laboratory testing procedure can be evaluated for its effect on the final result. Statistically the sum of the variances $(SD)^2$ of each step is equal to the variance of the entire procedure. $[(SD_{total})^2 = (SD_{analysis})^2 + (SD_{venipuncture})^2 + (SD_{biologic})^2 + \ldots]$ Since many of the variables have not been defined, it is imperative that the testing conditions be kept constant to maintain usefulness of the data produced by the procedure. In this discussion, the variables have been separated into laboratory and physiologic categories.

Determination and posting within the laboratory of *analytical* variation for each procedure performed is required by most laboratory certification programs. Analytical variation is expressed as $\bar{\chi} \pm 2$ SD of pooled specimens, using pools with two or more concentrations per test procedure. Although these values are determined from aliquots of a stable pooled specimen, they reflect only the analytical precision. The values are useful in determining the significance of changes in an analyte in a patient.

The analytical variance does not reflect the specimen handling or venipuncture steps. Specimen handling is subject to many problems, such as the following.

1. Contamination by the analyte per se: for example, plaster dust ($CaSO_4$) in the analysis of a specimen for calcium will result in a high value for calcium. The inadvertent use of a vacuum tube that contains EDTA for the collection of a blood specimen for alkaline phosphatase determination will yield a very low enzyme activity.
2. The length of time the serum or plasma is exposed to cells will effect the concentration of many analytes. The glucose concentrations in whole blood specimens in which glycolysis is not inhibited by fluoride or iodoacetate will decrease at about 4 mg/dl/hr in an unseparated blood specimen and much faster in a specimen with a high leukocyte count. Because most multichannel instruments analyze serum from blood collected and handled without glycolytic inhibitors, caution is suggested in the unrestricted extrapolation of data from studies of changes in serum values that can result from contact with cells.[2]
3. The temperature at which the specimen is held has a pronounced effect on some components. An excellent example is seen in cellular potassium distribution. Cooling a blood gas specimen on ice is effective in preserving the specimen for determination of *p*H and blood gas values. Cooling, however, causes potassium to leave the cells and results in a markedly increased plasma potassium level, making the specimen unacceptable for determination of potassium concentration. Serum lactic dehydrogenase is also less stable at reduced temperatures.
4. Errors in handling data may also occur, including transposition of digits in manually transcribed data. Failure to record and to apply the correct dilution factor to a value is a relatively common error occurring in the analysis of specimens that require dilution—for example, urine chemistry specimens or serum specimens with very high values.
5. Incorrect specimen identification can occur in the laboratory as well as in the venipuncture step and is addressed elsewhere in this book.
6. "*Short sample*" is used to describe an incomplete sampling of a specimen. It is a phenomenon almost limited to instruments with an automated pickup, especially those that use very small aliquots of specimen for analysis. Fibrin clots or "airlocks" are the most common cause: The former occur frequently in specimens from heparinized renal hemodialysis patients and require visual inspection and removal of the fibrin.

7. Evaporation of water from specimens waiting for analysis, especially those retained in small quantities, causes a significant concentration of the specimen. Such concentrating should be minimized by covering specimens that are not to be analyzed immediately.
8. Exposure to light will degrade many compounds, such as bilirubin, and should also be minimized.

VARIATION IN VENIPUNCTURE TECHNIQUE

Variation in venipuncture technique markedly affects the composition of the specimen. In the standard procedure, venous flow is occluded to allow puncture of the vein and quick removal of blood. This occlusion causes stasis and formation of a partial filtrate of blood. (The phenomenon has been used to determine ultrafilterable calcium in serum.) The length of occlusion varies for many reasons, including small veins and fat arms. Fortunately most blood specimens are obtained quickly and consist of small quantities of blood. It is not uncommon, however, to remove 50 ml of blood from one "stick," using several collection tubes, from patients undergoing special laboratory testing; patients with abnormalities in serum proteins are frequently in this category. In a series of experiments, we have demonstrated that the serum calcium of normal persons can increase from 9.7 to 10.1 mg/dl from the first to the fifth of consecutive 10-ml specimens drawn, contributing a variation of 0.4 of the total normal 1.2-mg range (8.9–10.1 mg/dl) in serum calcium.[3] Alterations in other analytes have also been observed owing to stasis. Studies on stasis without blood removal appears to yield different values than those described above that stimulate the conventional venipuncture technique.[4]

There are demands on clinical laboratories for increased specificity. Specificity of long used tests may suffer in this era of "polypharmacy." Spurious results may result from the presence of CRUD, an acronym coined by Seligson to indicate *C*ompounds *R*eacting *U*nfortunately as the *D*esired.[5] The interfering compounds are drugs and may be observed initially in procedures that furnish unsolicited data. Sulfasalazine (Azulfidine), a drug used to treat ulcerative colitis, is an acid–base indicator that is colorless at *p*H 10 and yellow–red at *p*H 12. Methods for creatinine determination based on the difference in absorbance at 510 nm of alkaline-picrate reacting material at *p*H 10.2 and 12.2 (Slot's method) are subject to a marked positive interference by therapeutic concentrations of sulfasalazine in serum. In addition, physiologically occurring substances may contribute to CRUD. Ketone bodies interfere with the alkaline-picrate methods used for determining creatinine concentration. The interference is method specific, with acetoacetic acid being more chromogenic than acetone and β hydroxybutyrate being nonreactive in three methods tested: Autoanalyzer, DuPont aca (a kinetic method), and a discrete adaptation of Slot's method.[3, 6] Without knowledge of the chemical procedure, it may not be possible to separate the chemical from the physiologic interference of a drug on the concentration of an analyte in blood or urine.

None of the above errors would be detected in the conventional quality-control schemes for laboratories, and for that reason the venipuncturist, specimen handler, and analyst should be acutely aware of these potential source of errors. The detailed procedure in the laboratory should include much of this information. Specific instructions for sample handling should be provided to all personnel involved in obtaining, processing, and analyzing clinical specimens.

PHYSIOLOGIC VARIABLES

Many variables in the testing procedure are physiologic. Some of them, such as age and sex, are classified easily and allow the relation of a value on a patient to an appropriate control or reference population. Others are more complex and not readily apparent or predictable owing to an influence of other factors, such as disease.

Posture of the person immediately before and during the venipuncture alters the distribution of body water and, thus, of some analytes. The prone plasma volume decreases when the patient stands. In normal persons, the average decrease is 12% at 15 minutes after standing; in sick persons, the plasma volume decreases as high as 30%.[7] The decreased plasma volume is due to the loss of plasma water from the vascular space. The concentration of filterable elements—sodium, potassium, glucose, urea—in blood does not change because these move out of the vascular space with the water, but those that are nonfilterable—formed blood elements, proteins, and analytes bound to protein—are retained at higher concentration in the reduced blood volume. This more concentrated blood is drawn at venipuncture. This phenomenon is usually ignored in setting normal values in the outpatient setting. It is commonly taken into consideration, however, in the normal values established for hospitalized patients. A recent study documents the effect of posture on the concentration of many common analytes in blood in normal adults.[8]

Many of the serum enzymes used for diagnosis of heart or liver abnormalities also occur in muscle. The activities of aspartate transaminase and creatine kinase are increased markedly after strenuous activity;[9] this increase occurs in both the conditioned athlete and the nonconditioned person. Not appreciating this fact when determining the normal range of enzyme activity could lead to a wider and less useful normal range and misinterpretation of the data—for example, on a patient with a suspected myocardial infarct.

The dietary influence on analyte concentration in body fluids is useful in the diet therapy of many diseases. The composition of the diet consumed can also alter the values obtained for many analytes in normal persons.

Serum uric acid level was determined in two studies at 7- and 10-year intervals.[10] In both, the normal range for serum uric acid was increased at 10 years in males and in one study in females. The changes were attributed to an increased protein intake in the persons studied, which shows that altered dietary habits over years can change analyte values and points to the need to reconfirm the normal values over time even for frequently ordered laboratory tests.

Many of the dietary influences on urinary composition may be eliminated by dietary abstinence from the component or its precursors after the evening meal and by collecting a timed urine specimen for analysis starting at 8:00 or 10:00 a.m. the following day.

A patient's sex can have a significant effect on the normal concentration of certain serum analytes. Males have higher normal concentrations of serum urea, creatinine, and uric acid: This difference is great enough to be reflected in the normal ranges used by many institutions. Lack of application of this fact would result in overlooking many abnormal values in female patients. A number of other analytes also exhibit sex differences, many of which readily become apparent only at sexual maturity. As the testing procedures become more standardized and with better analytical precision, more of these differences will become known. The clinical significance of the differences, however, remains to be determined in many instances.

The age of the patient also exerts marked influences on clinical laboratory values. As indicated earlier, the application of a test to a new patient population frequently causes problems when the new population is younger. A value of 0.9 mg of creatinine/dl of serum would be considered normal by most for either a male or female patient. Such a value, however, in a 1-year-old child, normal range <0.5 mg/dl, definitely is elevated. Interpretation of creatinine values obtained in young children is complicated by lack of sensitivity and specificity of most creatinine methods (*see* CRUD; ketones and creatinine).

Most physiologic functions decrease with age in adults.[11] A striking change is observed in glomerular filtration rate (GFR). Creatinine clearance, the most frequently used meausre of GFR, is related directly to age: Creatinine clearance ($ml/min/1.73m^2$) $= 133 - 0.64 \times$ age in years.[12] Normal ranges should reflect change with age.

Nearly all biologic phenomena exhibit rhythms. The most important to clinical testing is the circadian rhythm (*circa dies,* about a day). Some of these changes are due to rhythmic alterations in posture, activity, and eating, and, as discussed earlier, a change in posture can have a profound effect on the serum concentration of many analytes. Serum calcium is used as a key index in the diagnosis of hyperparathyroidism. The normal range determined under controlled conditions—early morning, fasting, in an ambulatory population—is 8.9 to 10.1 mg/dl of serum at the Mayo Clinic. The amplitude of the change over 24 hours for serum calcium was 0.5 mg/dl, or 40% of the normal range, whereas the mean range in serum phosphorus concentration over the same period equaled the entire normal range (4.8 mg/dl at 4:00 a.m. and 3.5 mg/dl serum at 8:00 a.m. to 10:00 a.m.).[13] Serum parathyroid hormone activity also exhibits a circadian rhythm.[13] For many diagnostic purposes, specimens must be obtained under standard conditions specifying the time of day.

The only rhythm that is routinely considered clinically is in the evaluation of adrenal function. Normal persons have a high a.m., low p.m. serum cortisol. Lack of this a.m./p.m. difference is diagnostic for Cushing's disease. Many investigators have shown that consumption of caffeine, consumed as coffee, will eliminate the rhythm in plasma cortisol.[14]

Coffee in the above context can be considered a drug. Many drugs are administered to alter physiologic functions. Many of these functions are monitored readily by changes in concentrations of analytes in body fluids.[15] The biochemical response of some individuals to drugs, however, confuses the diagnostic process.[16] Examples are elevated serum calcium and glucose concentrations seen in some patients receiving thiazide diuretics. The significant increases, up to 1.0 mg/dl of calcium and 50 mg/dl of glucose, are not uncommon in sensitive patients.

The physiologic variables noted above are not monitored in conventional quality-control procedures. The significance of these factors on the values obtained in clinical laboratory testing should demonstrate the importance of controlling and recording variables for individual patients.

The examples discussed are only a few of many that could be documented from individual experience and the literature. Many are routinely considered and controlled in the testing procedure. Others are only remembered when a "spurious" laboratory value or "laboratory error" is identified. They should serve to introduce the person involved in laboratory testing to an area of inquiry that will stimulate him to ask questions and identify and eliminate variables in the testing procedure.

REFERENCES

1. Roberts LB: The normal ranges, with statistical analysis for seventeen blood constituents. Clin Chim Acta 16:69, 1967
2. Laessig RH, Indriksons AA, Hassemer DJ, Paskey TA, Schwartz TH: Changes in serum chemical values as a result of prolonged contact with the clot. Am J Clin Pathol 66:598, 1976
3. Jones JD: Factors that affect clinical laboratory values. J Occup Med 22:316, 1980
4. Junge B, Hoffmeister H, Feddersen H–M, Röcker L: Standardisierung der Blutentnahme. Dtsch Med Wochenschr 103:260, 1978
5. Seligson D: Observations regarding laboratory instrumentation and screening analysis. In Benson ES, Strandjord PE (eds): Multiple Laboratory Screening, pp 87–114. New York, Academic Press, 1969
6. Slot C: Plasma creatinine determination. A new and specific Jaffe reaction method. Scand J Clin Lab Invest 17:381, 1965
7. Fawcett JC, Wynn V: Effects of posture on plasma volume and some blood constituents. J Clin Pathol 13:304, 1960
8. Felding P, Tryding N, Petersen PH, Hørder M: Effects of posture on concentrations of blood constituents in healthy adults: Practical application of blood specimen collection procedures recommended by the Scandinavian Committee on Reference Values. Scand J Clin Lab Invest 40:615, 1980
9. King SW, Statland BE, Savory J: The effect of a short burst of exercise on activity values of enzymes in sera of healthy young men. Clin Chem Acta 72:211, 1976
10. Griebsch A, Zöllner N: Normalwerte der Plasmaharnsäure in Süddeutschland. Z Klin Chem Klin Biochem 11:346, 1973
11. Masoro EJ, Bertrand H, Liepa G, Yu BP: Analysis and exploration of age-related changes in mammalian structure and function. Fed Proc 38:1956, 1979
12. Rowe JW, Andres R, Tobin JD et al: Age-adjusted standards for creatinine clearance. Ann Intern Med 84:567, 1976
13. Jubiz W, Canterbury JM, Riess E, Tyler FH: Circadian rhythm in serum parathyroid hormone concentration in human subjects. Correlation with serum calcium, phosphate, albumin, and growth hormone levels. J Clin Invest 51:2040, 1972
14. Avogaro P, Capri C, Pais M, Cazzolato G: Plasma and urine cortisol behavior and fat mobilization in man after coffee ingestion. Isr J Med Sci 9:114, 1973
15. Hansten PD: Drug Interactions. Philadelphia, Lea & Febiger, 1972
16. Young DS, Pestaner LC, Gibberman V: Effect of drugs on clinical laboratory tests. Clin Chem 21:1D, 1975

Patient and Specimen Identification: Future Considerations

Alvaro E. Pertuz

The Mayo Clinic is a large, integrated medical facility that renders care to both inpatients and outpatients. Almost all patients who come to the Mayo Clinic will have a phlebotomy during the course of their stay. This chapter deals with the method of identification using the latest technology: computer-printed bar code labels.

The Mayo Clinic uses a unit medical record. On his first visit, the patient is assigned a seven-digit number, which is known as his clinic number and which is used to identify both the patient and his medical record. This number is cross-referenced to patient name by use of the Soundex system and is never changed or reassigned. Given one combination of seven digits, there is only one Mayo Clinic patient that this can ever refer to. This number is the same whether the person is an outpatient or an inpatient (*see* Fig. 11-1A).

OUTPATIENT PHLEBOTOMY

When a patient registers as an outpatient, he is issued a "patient passport," which he must present at each area to obtain service. This passport has two computer-printed labels affixed to it: The larger label contains, in readable form, the patient's Mayo Clinic number, age, sex, address, civil status, and date of birth; the smaller label contains, in bar code form, the seven-digit Mayo Clinic number (Fig. 11-2). These labels are prepared at the same time the registration entry is made in the admission computer. As a result, the data in the computer as well as that on the labels cannot differ.

When the patient reports to the check-in desk for his phlebotomy, he is asked for his patient passport. The bar code label on the front is scanned by a light wand attached to a cathode-ray tube (CRT), part of the laboratory com-

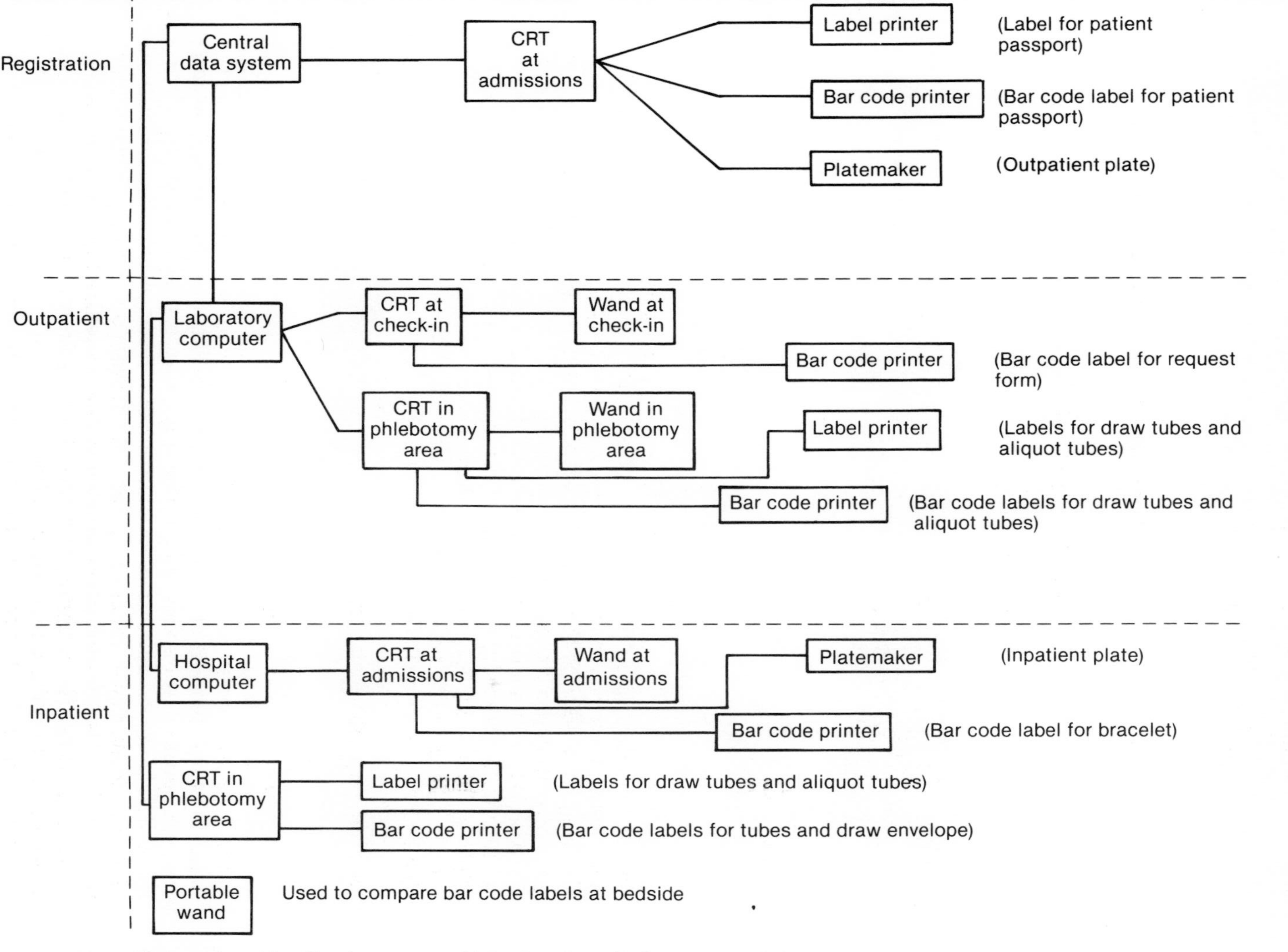

Fig. 11-1. *Mayo Clinic patient identification system.* (A) *Registration.* (B) *Outpatient.* (C) *Inpatient.*

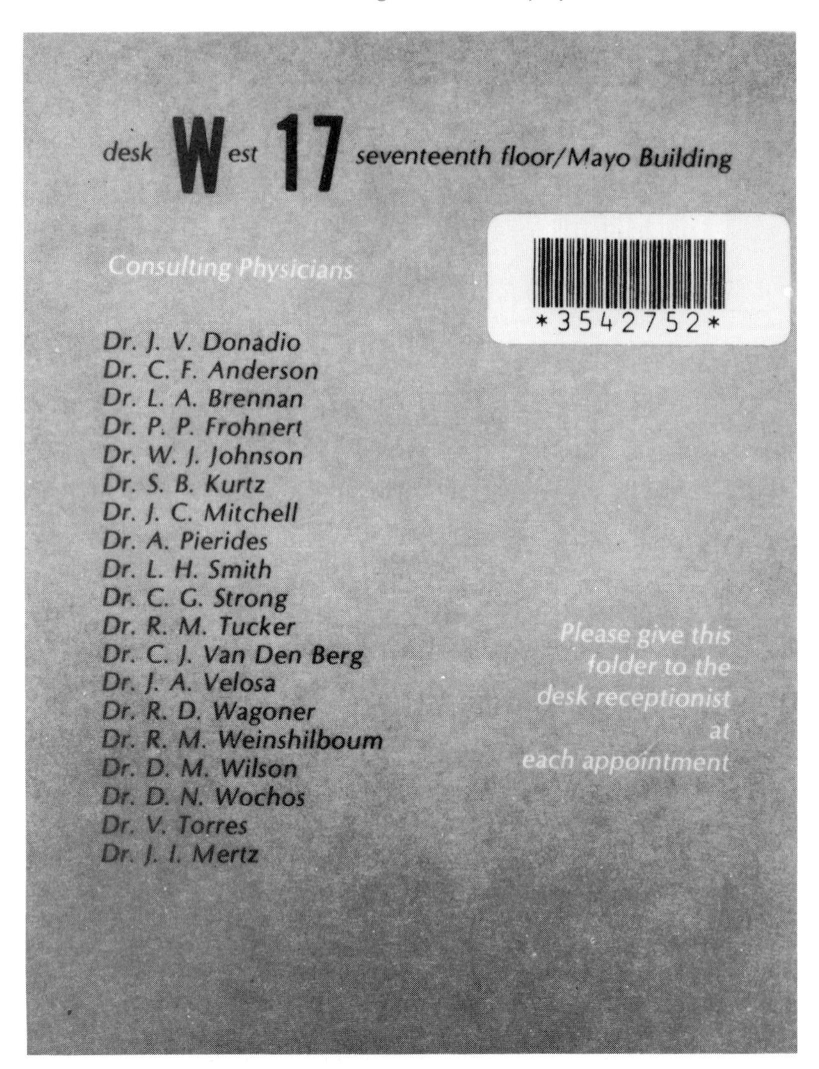

Fig. 11-2. *Patient passport with bar code label.*

puter. (The laboratory computer can communicate with the admissions computer to obtain demographic data.) The information read by the light wand (*i.e.*, the seven-digit Mayo Clinic number) is processed by the computer, and the CRT displays the demographic data available on that patient. The infomation is then compared with the label on the patient passport. This confirmation, as well as the visual confirmation of age (approximate) and sex, is part of the check-in process. The receptionist verifies verbally the patient's name and then requests that the computer system assign a five-digit accession number, which is displayed on the CRT. At the same time a bar code label is printed and affixed to the envelope that contains the physician-generated request form for blood tests. (This request envelope fits within the patient passport, which is designed as a pocket wallet and contains all the patients appointments.) The patient

Fig. 11-3. *Vacutainer with bar code label.*

passport is returned to the patient. The envelope that contains the physician's orders is kept by the receptionist and passed to the clerical personnel who support the phlebotomist.

On receipt of the request envelope, the clerical personnel use a light wand attached to a laboratory computer CRT to scan the bar code containing the accession number. The CRT displays the demographic information and the clerk compares the displayed information with the information imprinted on the order form. She then enters, on the CRT keyboard, a four-digit procedure number for each blood test the physician has requested. The system will print two labels for each draw tube and aliquot tube: One label will be a human readable label and will contain the Mayo Clinic number, patient's name, accession number, and an indication of what tests are to be drawn in that tube; the other label will contain, in bar code form, the five-digit accession number. These two labels are affixed on every draw tube. The labels that will be used for aliquot tubes are printed at this step but not used until the specimens reach the aliquot stations (*see* Fig. 11-1B).

The labeled draw tubes (Fig. 11-3) are passed to the venipuncturist. The information on the human readable label on the tube is verified with the patient, and then the venipuncture takes place.

The filled draw tubes and the labels for the aliquot tubes are passed to the aliquot station. Each aliquot tube is labeled with the human readable label (Mayo Clinic number, name, test, accession number) and a bar code label containing the five-digit accession number. Once filled, the aliquot tubes are delivered to the laboratory (*see* Fig. 11-4).

INPATIENT PHLEBOTOMY

When a patient is admitted to one of the hospitals affiliated with the Mayo Clinic, personnel at the hospital admission area request his patient passport. The bar code label on the front is scanned by a light wand attached to a CRT (*see* Fig. 11-1C). (The hospital computer can communicate with the central Mayo Clinic computer to obtain demographic data.) The information read by the light wand is processed by the computer, and the CRT displays the demographic data available on that patient. The data on the screen is verified with the patient, and admission personnel at the hospital will add other pertinent information

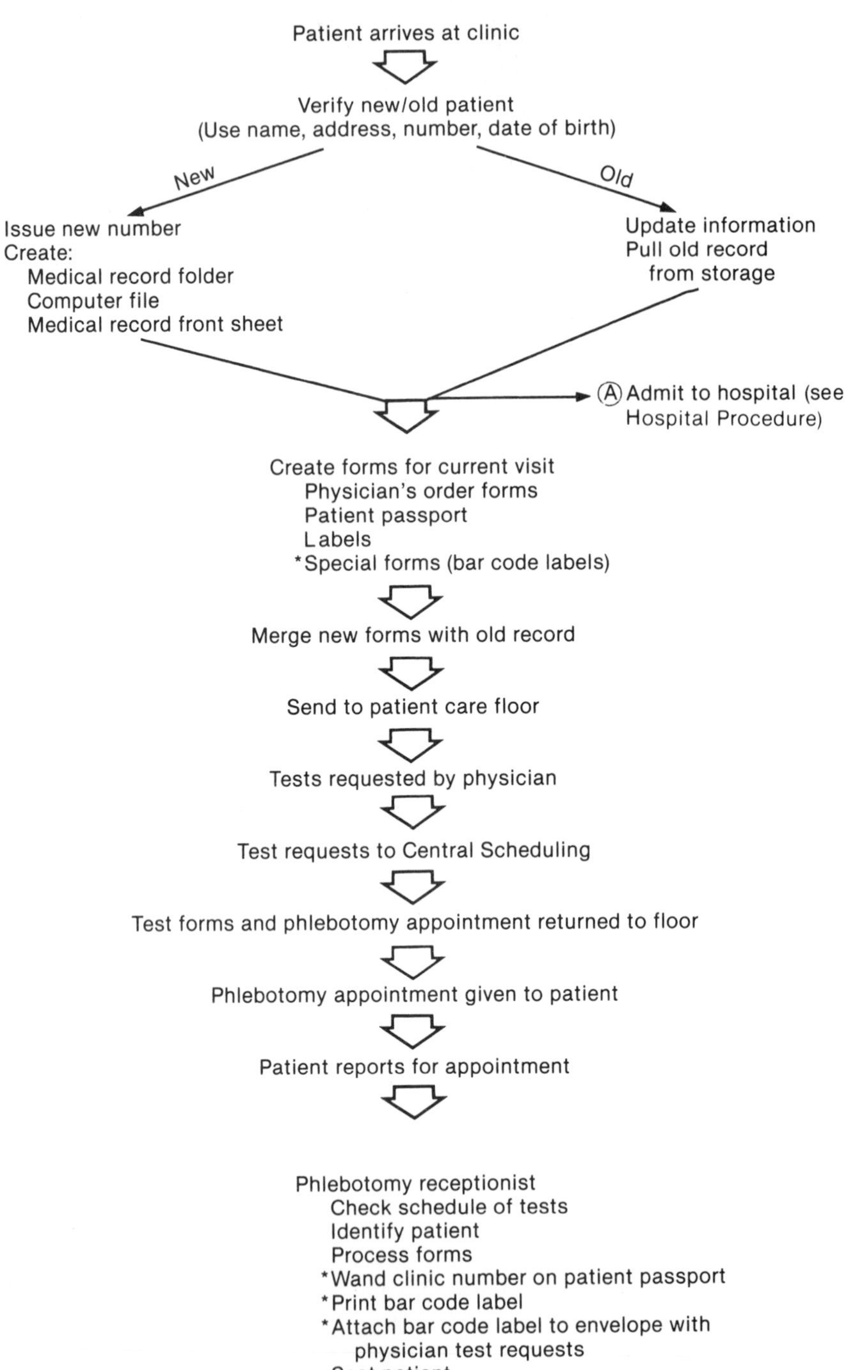

Phlebotomy receptionist
- Check schedule of tests
- Identify patient
- Process forms
- *Wand clinic number on patient passport
- *Print bar code label
- *Attach bar code label to envelope with physician test requests
- Seat patient

Room patient

⇩

Send envelope with requests to hospital venipuncture area

⇩

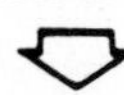

Venipuncture
- Enter name into laboratory computer
- *Wand accession number on request envelope
- Key in procedure numbers
- Print labels (human readable)
- *Print bar code labels (accession number)
- Label tubes
- Send labeled tubes and extra labels to Aliquot Station

⇩ ← (B) Specimen from hospital venipuncture area

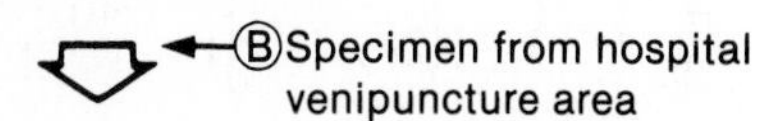

Aliquot Station
- Process specimens
- Aliquot specimens
- Place labels on aliquot tubes
- *Place bar code label on aliquot tubes
- Send specimens to laboratory

⇩

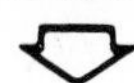

Laboratory
- Perform test
- Prepare report form
- Send report form to Central Distribution

⇩

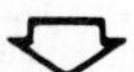

Central Distribution
- Separate report forms by clinic area
- Send to appropriate clinic

⇩

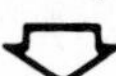

Clinic
- Sort by clinic number
- Place in appropriate medical record

Fig. 11-4. *Outpatient procedure.*

Fig. 11-5. *Plastic identification plate for hospital patient.*

Fig. 11-6. *"Draw envelope" with bar code label.*

EMERGENCY REQUESTS:

BLOOD TESTS			
Acid Base Balance, Venous	9102	APTT	9058
ABO and Rh Typing	9012	Carbon Monoxide	8649
Amylase, Serum	8352	Cholinesterase	8518
Bilirubin	8452	Differential	9120
Calcium	8432	Direct Coombs	9008
Chloride	8460	Drug Screen	8421
CK	8336	Indicate Drug:	
Creatinine	8472	Fibrinogen	8484
Glucose	8476	GPT	8362
GOT	8360	HLA Compatibility	9648
Hematology Group (CBC)	9109	LDH	8344
Ketone (Qual.)	8660	Metals Screen	8625
Lactate	8665	Phosphorus	8408
Potassium	8468	Platelets	9104
Prothrombin	9236	Protein, Total	8520
Sodium	8496	Reticulocytes	9108
Alcohol	8264	Rho (D) Immune	
Alkaline Phos.	8340	Glob. Crossmatch	8969
Ammonia	8388	Salicylate	8480
Antibody Identification Red Cell	8988	Sedimentation Rate	9152
Antibody Screen Red Cell	8956	Serum Osmolality	9340
		Urea	8492
		Uric Acid	8440

SAMPLE NUMBER(S)

MICROBIOLOGY CULTURES

Indicate culture desired by writing procedure number in box to the left of blood. No. ______

			Blood

PROCEDURE

2. Bacteria
5. Brucella
6. Fungi
11. Antimicrobial susceptibility
14. Neisseria

ADDITIONAL TESTS:

DATE TIME

REPORT AVAILABILITY

☐ MED. EMER. (< 1 hr.)

☐ PRIORITY TWO (ALL TESTS)
ORDER BY 6 P.M.: 3 HOURS OR LES[S]
ORDER AFTER 6 P.M.: 3 HOURS+

☐ PROCESS ROUTINELY
CHECK TESTS BELOW; WRITE IN IF NOT LISTED.

Only Tests Shaded Pager #______

COLLECTION PRIORITY
☐ IMMEDIATE (CALL HOSPITAL LAB)
☐ WITHIN 1 HR. ☐ TOMORROW A.M.
☐ FUTURE DATE:______

REDRAW REQUESTED:
☐ HEMOLYZED SAMPLE
☐ QUANT. NOT SUFF.
☐ WRONG TUBE
☐ MISSING TUBE
☐ SAMPLE NOT LABELED
☐ OTHER

LAB TESTS SUMMARY

MC1178/R 979

such as admitting date, admitting service, and room number. Once the data have been entered, the system will cause a computer-driven plate maker to prepare a plastic identification plate. The system will also activate a bar code printer that will print a bar code label containing the patient's seven-digit Mayo Clinic number. The plastic plate (Fig. 11-5) contains the patient's Mayo Clinic number, name, room number, hospital service, and other pertinent information and is used to imprint an insert for the patient's hospital identification bracelet. The bar code label is attached to the bracelet insert, and the bracelet is then assembled and placed on the patient's wrist.

Requests for blood tests are received in a central area at the hospital and

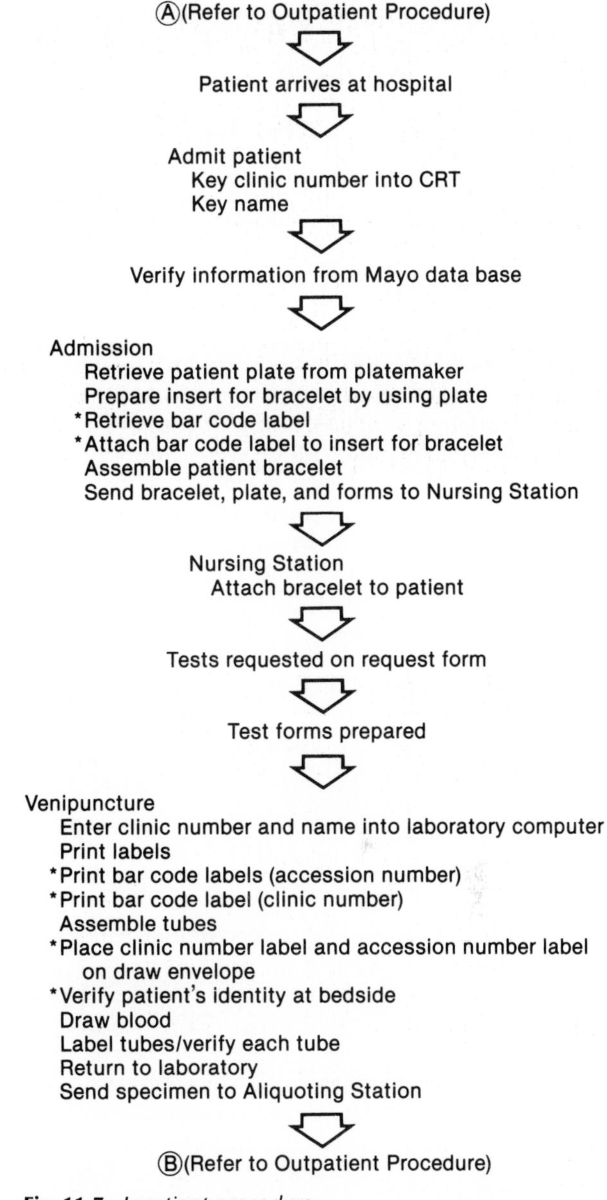

Fig. 11-7. *Inpatient procedure.*

are imprinted at the nursing station using the plastic plate that was prepared at the time of the patient's admission to the hospital. When the request is received, clerical personnel in the central phlebotomy area key the seven-digit Mayo Clinic number from the request form into the CRT. After entry, the CRT displays the name of the patient and other demographic data, which are verified with the printed information on the request form. The operator then enters into the CRT a four-digit procedure number for each blood test the physician has requested. The system will print two labels for each draw tube and aliquot tube. One label will be a human readable label and will contain the Mayo Clinic number, patient's name, accession number, and test names for that tube. The other label will contain, in bar code form, the five-digit accession number. The system will also print a bar code label that contains the patient's seven-digit Mayo Clinic number. The bar code label containing the Mayo Clinic number and one bar code label containing the accession are adhered to the front of a draw envelope (Fig. 11-6), which contains the requests for blood tests and the labels to be used for draw tubes. A series of draw envelopes organized by hospital area indicates the patients to be drawn during a specific round by a phlebotomist.

At the point the phlebotomist is to draw a sample (*i.e.*, at the patient's bedside), he will use a portable wand to scan the Mayo Clinic number bar code on the patient's bracelet and the Mayo Clinic number bar code on the draw envelope. The display unit on the wand will indicate a "go" message if the two numbers match. As blood is collected in each draw tube, the tube is labeled with two labels (carried in the draw envelope). One label is human readable, and one label contains, in bar code form, the five-digit accession number. As each draw tube is labeled, the portable wand is used to wand the accession label on the tube and on the draw envelope. The display unit will indicate a "go" message if the two numbers match (*see* Fig. 11-7).

The phlebotomist then returns to the central phlebotomy area, and the labeled draw tubes as well as the additional labels for the aliquot tubes are sent to the area where aliquots are prepared and labeled. Each aliquot that will contain a five-digit bar code label is then delivered to the laboratory for processing.

SPECIMEN IDENTIFICATION

1. All samples are drawn in Vacutainer tubes. In some cases a syringe is used, but the sample is then transferred to a Vacutainer tube.
2. A standard 12-mg aliquot tube is used for almost all aliquots.
3. The human readable label is 1 $\frac{11}{16}$ inches long by 1 $\frac{1}{8}$ inches high and contains six lines of information with up to 16 characters each.
4. The bar code label is 1 $\frac{1}{2}$ inches long by $\frac{5}{8}$ inches high and contains up to seven-digit number (in bar code form) with a start-and-stop code.
5. The Mayo Clinic is using high-density code 39 as its standard bar code.
6. All hardware is available commercially and can be used without modifications. The portable wand requires custom software, which was provided by the manufacturer.

Part V

TRAINING AND ADMINISTRATIVE CONSIDERATIONS

12

Training the Phlebotomist

Venipuncture

Eppie McFarland

A career as a phlebotomist is today an option for the high school graduate. Many institutions are now either training people on the job or establishing a training course within an educational center. Phlebotomists, in essence, represent the laboratory and are the laboratory's contact with patients, nurses, and physicians. Thus phlebotomists must be trained to approach the patients confidently and to obtain specimens quickly, properly, and without undue discomfort to the patient.

SELECTION OF CANDIDATES FOR TRAINING

Venipuncture is a procedure requiring both knowledge and skill to perform. In a survey at our institution, our phlebotomy supervisors listed the following important characteristics of a good venipuncturist.

1. Accuracy.
2. Mental alertness, quick and clear thinking, and attention to detail.
3. Conscientiousness and responsibility.
4. Ability to follow instructions and take corrections.
5. Manual dexterity and speed. Good number recall, recognition.
6. Pleasant personality and ability to cooperate with coworkers and hospital staff. Empathy and friendliness to patients.

Considerations in Selecting Applicants for Interviews

1. Good high school grades. Especially note if applicant has reading problems, important because of volume of names and numbers dealt with.
2. Dependability as evidenced by high school attendance. Frequent absences from school will usually continue into the job.

3. Good comments from employment or school counselor.
4. Satisfactory work references from previous employer.
5. Good health record. Important not to have foot or back problems. Phlebotomists do a lot of walking and bending in an 8-hour shift.
6. Willingness to work different shifts, weekends, and holidays. Most health care centers offer 24-hour medical services, 365 days a year; they thus have paramedical people on duty to cover these hours.

Characteristics to Look for When Conducting Interviews

1. Neat appearance. Neatness and good grooming are extremely important because phlebotomists serve as representatives of their laboratory. A simple controlled hair style and a moderate amount of makeup and jewelry are appropriate. Long hair should always be tied or pinned back at work. Well-manicured nails are also important because the phlebotomist's hands are highly visible during the venipuncture procedure. Polish, if worn, should be a conservative color. Clothing and shoes must be clean and conservative. If one's institution has a dress code, this should be reviewed with the applicant at the interview.
2. Poise (composure, relaxed nature)
3. Alertness (comprehension), prompt responses
4. Pleasant personality (likability, charm, friendliness)
5. Maturity
6. Initiative
7. Ability to speak well
8. Enthusiasm and sincerity
9. Cooperation (positive attitude, eagerness to please)

OBJECTIVES OF TRAINING PROGRAM

The first step in planning any type of training program is to establish objectives that indicate the kind of behavior expected of the trainee. Objectives of a phlebotomy training program should be those that allow the trainee to gain confidence and professionalism in the art of obtaining biologically representative sample for the laboratory to analyze. Laboratory studies have shown that the way blood is drawn, in what order tubes are drawn, and how long tourniquets are left on greatly affects test results.

After training, phlebotomy trainees should be able to do the following.

1. Explain the physical layout of the area they will be working in.
2. Identify the various pieces of equipment used in the venipuncture procedure and describe how they are used.
3. Identify the various forms associated with the venipuncture procedure and explain the purpose of each form.
4. Make proper identification of patients before obtaining blood samples.
5. Know the amounts of blood and types of tubes needed to perform each test.
6. Identify and locate the various veins used in venipuncture.
7. Perform the venipuncture professionally without causing undue discomfort to the patient.

8. Know special procedures such as how to
   - label blood bank samples properly to comply with AABB standards.
   - draw bloods using isolation techniques.
   - draw bloods using sterile technique for blood cultures.
   - draw bloods from patients who have cannulas, fistulas, and intravenous lines.
9. Collect microsamples for various chemistry and hematology tests on children as well as on adults with difficult veins.
10. Make peripheral blood, reticulocyte, and malaria smears.
11. Perform the Duke bleeding time test.
12. Gain the patient's confidence by appearing professional and skillful in the art of phlebotomy.
13. Gain the physician's respect by being cooperative and by giving prompt service.
14. Gain job satisfaction as their expertise in the art of phlebotomy increases.

TRAINING PROGRAM

The training program should comprise both didactic instruction and empirical training.

Didactic Instruction

Trainees should receive a minimum of 12 hours of didactic instruction, which will help them to understand the important steps to follow when obtaining a blood specimen. Ideally this didactic instruction should be given by the instructor, but experienced phlebotomists may help with some of the training. Trainees thus will be trained consistently by working with the same people.

All persons involved in training should be given an outline of the procedures to follow. A training manual is important so that all procedures are well documented. Slide tape presentations describing venipuncture procedures are ideal so that after the trainee has received instruction from the trainer, he can sit down quietly and review these procedures again.

A designated amount of time should be spent on each of the following.

1. Patient Identification
   This important procedure should be explained carefully to the trainee. Incorrect identification could result in a patient's being transfused with incompatible blood, having a reaction, and dying, which in turn could result in a law suit for the hospital. Any bloods drawn on the wrong patient could result in the physician making an incorrect diagnosis or giving wrong treatment.

   A good procedure for identification of the patient is as follows: In a hospital setting, the patient's name and identification number on the request form and envelope should be checked on the door card. These forms should then be taken into the room. The phlebotomist should identify herself to the patient, stating that she is from the laboratory and is going to take a blood sample. She should then ask the patient to tell her his name and, if it is unusual, to spell it for her. The information on the request forms and envelope should be compared with the patient's identification band. If the patient has no identification band, the ward clerk or nurse should be asked to identify the patient for the phlebotomist, and an identification band should be put on the patient's wrist.

If a discrepancy occurs in the identification, the blood sample should not *be drawn until identification has been established.*

If the patient is incoherent, a relative or friend may identify the patient. If no one is present, the nurse may make positive identification. In an outpatient area, the patient should be asked his name and, if unusual, to spell it. The name should then be compared with that on the request forms; if everything matches, blood can be drawn.

2. Patient Relationship
   The manner in which the phlebotomist approaches the patient may have a direct bearing on the patient's response to venipuncture. The trainee should learn the following.

   - To be friendly and tactful upon entering the patient's room
   - To introduce herself to the patient and tell him why she is there and what she is going to do
   - To speak quietly at all times. If the patient wants to talk, she should listen, but not to the extent of disrupting work. If the patient discusses personal information, the phlebotomist should keep it confidential
   - To respect the religious beliefs of the patient. If the hospital or clinic services a particular ethnic group, it may be helpful to learn a few sentences of that language or to have some cards with specific phrases in several languages so that the phlebotomist can let the patient know what she is going to do.

If you continue to see the same patient frequently, the phlebotomist should become familiar with his interests, hobbies, or family and use these as topics of conversation. Many patients in the hospital are lonely and need a friend. Occasionally an extremely ill patient does not wish to talk; his wishes should be respected.

One should remain pleasant even if the patient is disagreeable. Many times the patient is afraid of what might happen to him; if he is alone, his fear becomes even worse. The phlebotomist should be firm in this situation, remain cheerful, be kind, and do her job confidently.

Before collecting blood on a child, one must establish rapport with both the child and the parent (if present). The child should be greeted in a friendly, confident voice. The phlebotomist should explain to the child what she is going to do; small talk sometimes helps to put the child at ease. Young children who do not understand words are usually pacified by the sound of a confident voice.

An older child usually is easier to deal with; he should be told what is going to happen and can be asked to help, such as in letting the phlebotomist know when blood is entering the tube or by holding the gauze or Band-Aid.

A child should never *be told that it will not hurt but instead, that it will hurt a little bit and that if he remains still, the test will go much faster. After the test has been finished, the child should be given some type of reward, such as an animal badge for smaller children or a small badge of courage with a positive message such as "Best Behavior Award" or "I was a good girl/boy" for older children. By spending a little extra time boosting the morale of a child and giving him a little extra TLC (tender loving care), the phlebotomist will make the job easier for herself or for one of your fellow phlebotomists the next time the child needs to have a blood test.*

Before leaving the patient's room, one should check to see that everything has been returned to the laboratory tray.

3. Forms
The trainee should become familiar with all forms that need to be completed for venipuncture, including the physician's test ordering form and the forms, envelopes, and labels that will be used by the phlebotomy team.

4. Accession Order
Each request for a blood specimen must be accessioned to identify all paperwork and supplies associated with each patient. The trainee must learn the function well because it provides the laboratories with demographic data on the patient, such as full name, patient identification number, address, birthdate, sex, the physician's name, and location of the patient.

5. Equipment Needed to Perform Venipuncture
The instructor should display different sized needles, syringes, evacuated tubes, and tourniquets. All equipment to be used should be reviewed and demonstrated by the instructor and tried by the trainee. This "show-and-tell" method of training is very effective.

6. Vein Anatomy
The trainee should be given some background on veins in the forearm, hand, and foot. The use of vein charts and perhaps a lecture by a staff physician explaining the circulatory system would be beneficial.

7. Basic Steps in Performing Venipuncture
To assure that samples are collected in the same manner by everyone, the National Committee for Clinical Laboratory Standards (NCCLS) approved standard for the collection of diagnostic blood specimens should be used for training.[1] *This guide can serve as a reading assignment for the trainee and should then be reviewed with the instructor. Each step should be discussed in detail. During the training period, the trainees should be encouraged to ask questions about anything they do not understand.*

8. Simulated Patient Arm
The simulated arm can be constructed of materials ordinarily found in a clinic laboratory. The procedure we use for construction of the simulated arm is a modification of the simulated arm as described by Taaca.[2] *The advantages of using the simulated arm are that the trainee can become familiar with assembling the equipment, applying the tourniquet, feeling the vein, pushing the evacuated tube onto the needle, removing one tube and inserting another, and using a syringe and needle. Another advantage of the simulated arm is that it cannot be pulled away in protest of pain as a live patient might do. Thus, the student is kept from becoming "scared stiff" during the first hours of introductory phlebotomy practice, and instead can gain self-confidence and ease in performing the procedure.*

9. Terminology
The trainee should be provided with a list of medical abbreviations, laboratory terminology, and meanings. If the laboratory has test profiles, the various tests offered in each profile should be listed. The trainee shuld be taught to become familiar with the different types of patient classifications, such as cardiology, neurology, hematology, and oncology.

10. Physical Layout of Location Where Phlebotomist Will be Working
    If the trainee is going to work in a hospital and is being trained in another area, it would be ideal to have a map of the various floor plans so that the emergency rooms, intensive-care units, and pediatric floors, among others, can be pointed out. When the trainee goes to the hospital for her final training period, the areas can then be pointed out in more detail.

Empirical Training

Observation and Assistance Phase. The purpose of this phase is to give the trainee a chance to become more familiar with procedures, equipment, and types of patients. It also gives her the opportunity to become familiar with occasional problems, such as a patient becoming ill, fainting, or having seizures and the best ways to deal with these developments.

The trainee should observe the entire venipuncture procedure and assist the instructor by performing the following tasks.

1. Identifying the patient
2. Assembling supplies and labeling tubes
3. Applying the tourniquet
4. Cleansing and selecting the vein site
5. Obtaining samples using either evacuated or syringe method
6. Bandaging arm and dismissing patient
7. Obtaining a blood culture specimen
8. Handling a hepatitis sample

Performance Phase

This basic phase gives the trainee the opportunity to develop the confidence and experience needed to become an expert phlebotomist. While working under the supervision of an instructor or senior technician, the trainee should attempt venipuncture on a patient with good veins and who is having only a small amount of blood drawn. If she experiences difficulty, the instructor should take over and complete the procedure. After the patient has been dismissed, the trainee should be given an explanation of why she did not obtain the sample and some helpful hints of what to do the next time. After the trainee has performed her first venipuncture, she should progress to obtaining more samples using both the evacuated and syringe methods on patients with good veins. When the trainee has confidence, she should attempt patients with difficult veins.

After about 1 week of close supervision, she should be allowed to work alone but should ask for assistance if the sample is not obtained after two attempts (or if after one try she is unable to locate another vein). After the trainee has worked for 15 days, she should be able to draw blood consistently on eight to ten patients an hour. Speed is not the important factor at this time—accuracy is—but speed will come with experience. After the trainee has been doing venipunctures for 3 weeks with someone close at hand to assist her, she should be ready for more intensive special training.

The trainee should accompany the assigned senior phlebotomist for 3 more weeks and learn the following.

1. All special procedures in isolation units and intensive-care units.
2. Microtechnique in pediatric units and pediatric intensive-care units
3. Slide-making
4. Duke bleeding time tests
5. Emergency-room procedures
6. Diabetic urine tests
7. Telephone work, such as answering questions and taking orders from physicians
8. The order entry function and the obtaining of labels
9. Procedure for proper transportation of specimens
10. Evening shift procedures
11. Night shift procedures

After training in all of the above, the trainee should be allowed to go out on her own. She should be given an easy floor on which to start so that her confidence will build.

Most trainees are trained sufficiently in 6 weeks and feel quite confident after 3 months on the job.

Progress Evaluation

An evaluation sheet will list all procedures that the trainee should perform satisfactorily. This sheet enables the instructor to monitor the trainee's progress and to know how effective the training program was. If a trainee is not doing well in a specific area from the progress evaluation, the material should be reviewed with that person. I have found this to be an effective way of training. Each person who assists the trainee should be asked to evaluate the trainee on the following items.

Patient identification
Attitude
Interest
Rapport with patient
Maturity
Neatness
Punctuality
Attendance
Venipuncture technique
Syringe technique
Blood culture technique
Isolation technique
Capillary technique
Slide-making technique

During the training period, a number of tests should be given daily or weekly to the trainee, either oral or written quizzes. These tests will give a measure of how well the student has learned and how effective the training has been.

After the trainee has been working for 3 months, a comprehensive examination should be administered on all aspects of her work. This gives a better record of her daily progress and how well she has assimilated all the procedures and practices of phlebotomy.

Remember, if the learner has not learned, the teacher has not taught.[3]

We have used this method of on-the-job teaching of trainees for the past 2½ years and have been pleased with the results. The training program has provided us with well-trained, efficient, and independent ambassadors of the laboratory.

REFERENCES

1. National Committee for Clinical Laboratory Standards (NCCLS): Standard Procedures for the Collection of Diagnostic Blood Specimens by Venipuncture. Adopted Standard: H3-A_5. Villanova, NCCLS, 1980
2. Taaca EB: Simulated patient "arm" help in venipuncture instruction. Lab Med 6(12) p 40–41. December 1975
3. War Manpower Commission: The Training Within Industry Report, p 195. Washington DC, Bureau of Training, Training Within Industry Service, War Manpower Commission, 1945

Arterial Puncture

Susan Stumpf

The personal qualities of an arterial phlebotomist should include maturity, conscientiousness, reliability, dexterity, and discipline. Some scientific background is required because she must be able to take data from various forms of respiratory apparati. The phlebotomist must also be able to communicate well with patients and manipulate them successfully.

The time allotted for training personnel in the skills of arterial puncture varies with each facility and the daily volume of work. The training time required by each person depends on past experience and that person's aptitude. I recommend training phlebotomy personnel for arterial punctures only after they are comfortable with venipunctures. They should *not* be trained for both at the same time.

The first step is to give trainees copies of arterial puncture procedures. They should read the procedures thoroughly and schedule a meeting to discuss them. The instructor should allow about 2 to 3 hours to discuss the procedures step-by-step and explain the rationale and anatomy involved in arterial punctures. The discipline guidelines and techniques needed for this procedure should be established at this time.

Before the lecture begins, all equipment used for obtaining arterial samples should be assembled, including syringes, anticoagulant, needles, Luer tips or some dead-ending device, gauze, alcohol wipes, identification bands, examples of charting data sheets, mock-up arterial cannulation systems (including the flush system), dialysis cannula sampling parts (if used), Swan–Ganz catheter system, and a practice arm for techniques. The equipment should be demonstrated as techniques are being discussed. The trainee should then perform the same procedure immediately afterwards. This is a very valuable technique and makes the first few bedside experiences go much smoother.

POINTS OF EMPHASIS

Throughout the lectures, certain topics should be emphasized, explained, and understood for good performance of the task. When explaining the anatomy involved with arterial punctures, for example, the instructor should stress that the trainees know and understand the anatomy so that they can function in a crisis and prevent damage to the patient. The points to mention on individual arteries (Fig. 12-1) include its depth, the location of nearby nerves, collateral systems, and the chance of hemorrhage. With brachial arteries, the possibility of accidentally puncturing a vein with an intravenous (i.v.) insert should be stressed. This blood is easily confused with arterial blood because the venous blood is diluted by the i.v. solution and looks bright red. Blood gas analysis will not be helpful.

When discussing the femoral artery, the instructor should stress the depth of the vessel and possible hemorrhage into the muscle tissue. Moreover, there is a danger of dislodging a plaque in the artery with the needle. Other problems include arteriovenous fistulas, emboli, pain, closeness of nerve, and danger of infection.

Another point of emphasis is heparin dilution of arterial samples. The problem of too much heparin in the syringe should be eliminated if the procedure is followed. The biggest problem is trying to "get by" with a short sample when the heparin to blood ratio is quite high. Heparin dilution will cause falsely lowered Po_2, Pco_2, *p*H, and hemoglobin values. Quite often the sample may be clotted in the case of a difficult stick. A consistent volume drawn for each sample will control this problem. About 2 ml to 3 ml of blood works best for us.

The arteries will spasm occasionally because of an irritated arterial wall, and the artery will shut for 10 to 20 minutes. Quite often trainees with shaky hands and needles will induce a spasm. The best thing to do is to withdraw the needle and go to another site.

Many trainees express concern that their act is causing great discomfort to the patient. The trainee should handle this feeling by the use of psychology and by learning the task well. If she does hit a nerve accidentally, the patient will experience pain but will not, of course, die.

Another problem is that of applying pressure to the puncture site. Discipline is critical. If the phlebotomist is not to be responsible for holding the puncture site, the shifting of responsibility should be defined clearly. Whoever holds the puncture site must be instructed to apply pressure for 5 minutes (minimum) by the clock—10 minutes for a femoral artery. While applying pressure, the person should not obstruct the flow of blood within the artery. One should be able to feel the pulse while applying pressure.

If another person offers to hold the pressure on the site, they should be told how many minutes remain. Pressure bandages or Band-Aids are not al-

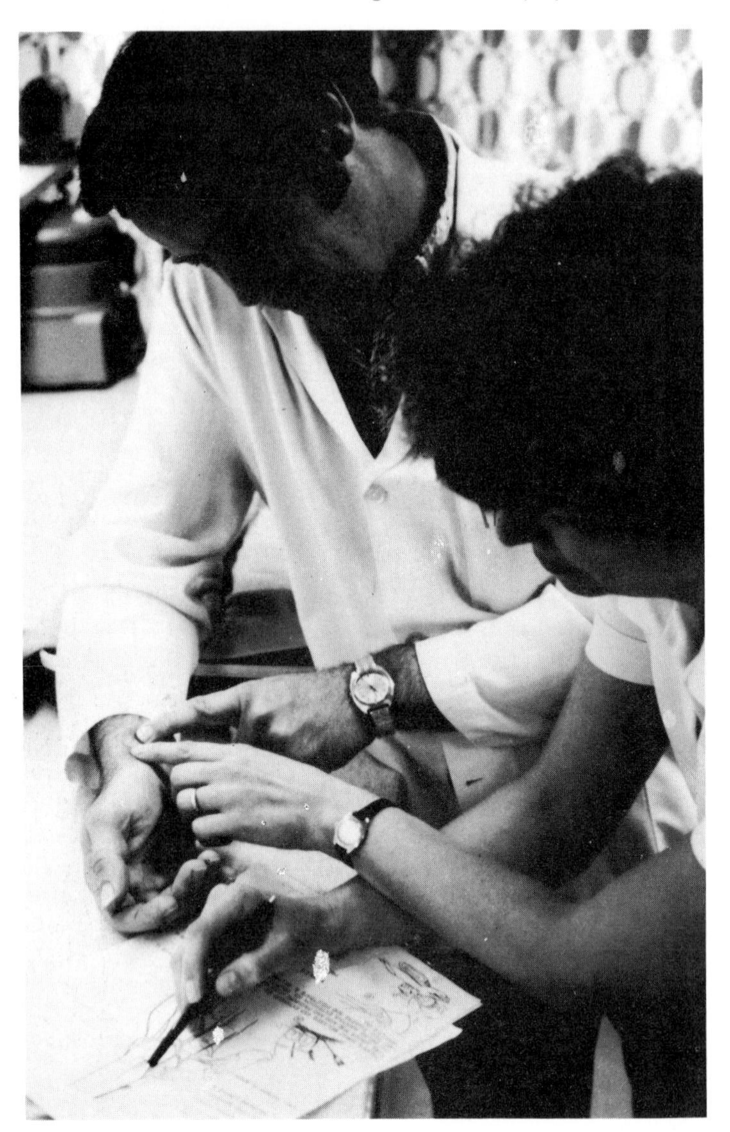

Fig. 12-1. *Teaching radial artery anatomy.*

lowed on a routine puncture site: They tend to mask any bleeding and are not effective in applying pressure. The pulse downstream from the puncture site should be checked once the bleeding has stopped to confirm continued blood flow.

The patient's body temperature must be accurate. Ideally the temperature should be measured at the time of sampling. Axillary temperatures are not acceptable because of variable perfusion rates in patients.

Data collection is critical for interpreting the blood gas analysis. The phlebotomist should be alerted to abnormal breathing patterns such as apnea spells and Cheyne–Stokes and Kussmaul breathing. Patient positions—prone, supine, sitting—are important.

Another point of emphasis is the icing of a blood gas sample. The rationale behind this is that icing (0–5°C) will slow down the metabolism of the white blood cells to preserve the level of oxygen that existed when the sample was drawn. This is not sufficient enough for samples with extremely elevated white blood cell counts—for example, leukemic patients or septic patients. These samples should be iced immediately and rushed to the laboratory, the staff alerted immediately to the problem, and the sample analyzed quickly. Some physicians have even had patients come into the laboratory to have their blood samples analyzed immediately.

Another point of concern with blood gas samples is the time lag from sampling to analysis. It is poor practice to go on a sample run and collect more than one or two samples at a time. The policy at the Mayo Clinic allows 45 minutes maximum (on ice) from sampling to analysis; most samples are analyzed and reported with 15 minutes. If the sample is to be split for other tests, it should be mixed well and aliquoted immediately before icing. Icing in particular will cause the red blood cells to release potassium when they become cold.

APPROACH TO PATIENTS

The psychology of approaching the patient can be critical to the validity of the sample itself. This is an important topic to discuss with trainees: It is easy to become involved with the mechanics of the procedure and forget the patient.

Many patients are frightened. They often feel as if they are suffocating; adding to their stress is harmful and unnecessary.

1. Read your patient! Analyze the total situation and gear your approach accordingly.
2. Explain what you are about to do in layman's terms.
3. Keep needles out of sight as much as possible, especially with children.
4. Be prepared. Have everything ready to go before you enter the patient's room.
5. Act quickly and with confidence.
6. Be firm but gentle (not wishy-washy).
7. Develop skills well.
8. Check to see that the patient has been prepared properly. Check that he is resting in steady state and is not eating or walking around. Check to see that he has had his mode of oxygen delivery on for the prescribed amount of time and that there has been no suctioning, ventilator changes, or chest physical therapy or intermittent positive pressure breathing (IPPB) treatments given within 30 minutes. This will save the embarrassment of sticking the patient needlessly.
9. On return visits to patients, remember they know it hurts.
10. If the patient is becoming extremely agitated, let him rest (depending on the circumstances); if the sample is hard to draw, another person should attempt to harvest the sample. Crying and hyperventilation (breathing at an abnormally fast pace) will change the blood gas values significantly

within seconds. The P_{CO_2} will be lowered and the P_{O_2} will (normally) be elevated slightly. Likewise, breath-holding may cause the P_{CO_2} to become elevated, and the P_{O_2} will be lowered significantly.

11. Give the patient some TLC (tender loving care). Be empathetic. Touch. Listen (within reason). Be gentle. Do not be nervous. Do not show negative feelings that will frighten the patient or his relatives. The time spent applying pressure to the puncture site is ideal for holding the patient's hand and making him more comfortable. The patient needs friends. Even patients who are intubated need someone who cares, although they cannot converse. Do not dwell on the negative; instead, emphasize the positive.
12. If the patient is not as the physician ordered, verify the order and contact the nurse to see whether the order has been carried out and set up a mutual time to return for sampling.
13. If something is functioning improperly or the patient is distressed, ask a respiratory therapist, nurse anesthetist, or doctor to check on the patient.

THE PRACTICE ARM

The practice arm (Fig. 12-2) allows the trainee to work out the strategy of performing an arterial puncture. Trainees should talk to the arm as if it were a patient. It helps them to verbalize their approach and work out the placement of equipment.

The practice arm, purchased or homemade, is not exactly like a person, but it is a good starting place and is helpful when practicing different ways of holding the syringe. Also, if someone is having technique problems, the practice arm is helpful.

VENTILATORS, ANALYZERS, AND VITAL STATISTICS

Learning to take readings from the ventilators and other apparatus is perhaps the hardest skill to learn. The best way to learn is to get a manual, study it, and review all the required readings with a respiratory therapist familiar with the machine.

An oxygen analyzer should be used to verify the percentage of oxygen in a closed system—that is, a ventilator. The analyzed gas should be humidified and the analyzer calibrated daily. Repetition is the best teacher.

THE REAL ARM

The trainee is now ready for patient contact. She should observe three or four situations before attempting one. The trainee should have observed several different techniques throughout the training period.

If possible, a patient with a good pulse who is somewhat unresponsive and in a no-crisis situation should be chosen. The trainee should be talked through the entire procedure step-by-step. The instructor may hold the patient's hand and arm, which will allow him to be nearby and prepared to take over at any time.

After the first few arterial punctures, the trainee may observe other arterial phlebotomists for different techniques.

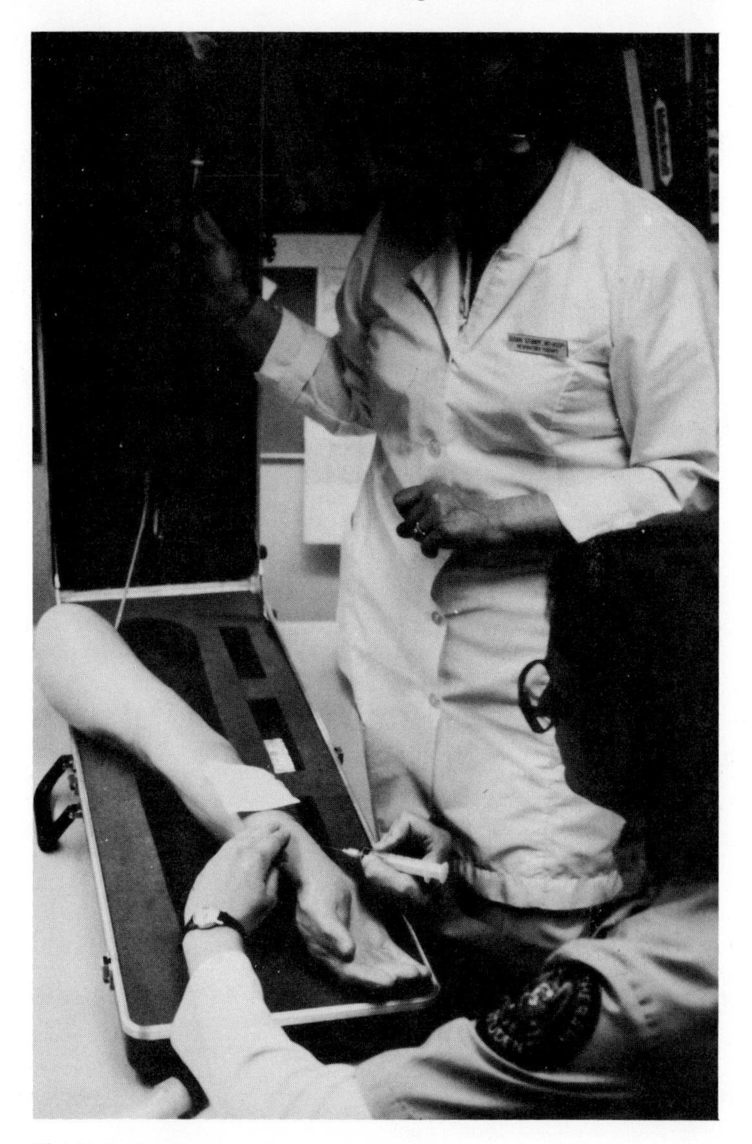

Fig. 12-2. *Student training on practice arm.*

EVALUATION AND CRITIQUE

The instructor should be liberal with praise, should make suggestions, and should allow the trainee to experiment on the practice arm. About 3 or 4 days (five to ten sticks) of practice sticks should be allowed before the instructor does preliminary evaluations. After a few weeks, the trainee should be evaluated by observation and then be allowed to venture out on her own. It is important to maintain good communications with the trainee and discuss problems daily. Once the trainee has completed all phases and evaluations, she is then certified by the physician in charge.

13

Supply Management and Quality Control

Betty Winkler

The management and quality control of supplies can be simplified by implementing a program of usage accounting and sample tube testing. These procedures can be performed regularly by either a large or small venipuncture department. Annual inventory census and sample tube testing along with a reliable vendor and a contract for major supplies can contribute to the smooth flow of supplies through the department. Purchasing and quality control of these supplies are interdependent. In practice, they are separate functions and will be discussed that way for reasons of clarity.

INVENTORY CENSUS

Annual inventory census helps to communicate use of the evacuated tube and the needle to the vendor and supplier. Once each year usage counts should be made for a period of 1 month. From these figures, one can estimate annual use. These estimates are useful to the stockroom, the vendor, the supplier, and the purchasing department. They help the vendor to maintain an adequate supply and aid the supplier in estimating production needs.

AGED TUBES

At least a 2-week supply of tubes and needles should be kept in the venipuncture department at all times. Rotation of this stock is essential upon receipt of every new order in accord with the expiration date. The vacuum draw of the evacuated tube may decrease with age, which could cause an adverse test result. Tubes made from soda lime glass have a shorter stable life span than those made from borosilicate glass. This breakdown of the soda lime glass tube could affect some chemical test results. If tubes made from soda lime glass are used, they should be checked for trace metal (calcium and magnesium) contamination beforehand. All brands of evacuated tubes are coated internally with a very thin layer of silicone to prevent cell hang-up on the sides of the tubes during centrifugation. Silicone can also break down with age, resulting in poor specimens after centrifugation.

The importance of observing the manufacturer's expiration date, clearly printed on every container or box, should be stressed again. As a practical matter, it is well to set up stock control and ordering procedures such that only a sufficient number of tubes are in stock to take care of a department's needs for a reasonable period. That time should allow for delivery of new materials. The crisis situation of running out of tubes can be avoided this way, and one can be assured that the tubes will be used well before expiration date.

Overstocking should be avoided for two reasons: Tubes are perishable, and tubes are expensive. Too much on the shelf can result in a loss of interest on the dollar. Attention to these concerns will be given if the responsibility for ordering and stock rotation is delegated to a lead technician or an assistant supervisor. If no one person is identified as being responsible, the task becomes everyone's and therefore no one's, and it is quite likely that it will not get done.

NEEDLES

The venipuncturist's most valuable tool, other than her skillfully trained hands, is the needle. The phlebotomist is always alert to defects and problems with needles and should be listened to carefully for any criticism of the needle. Because this needle is used to perform the skill that the phlebotomist prizes and has developed through time and experience, it is a tool that should be selected with the phlebotomist's input. Asking the venipuncturist to perform such a delicate procedure with a needle of poor quality would be unfair.

Many phlebotomists have found that they have better control when performing the venipuncture if the shorter 1-inch needle is used rather than the longer 1½-inch standard needle, which they feel is more awkward. The 1-inch needle is available for both the single sample and the multisample collection systems. It is not a standard needle for some brands but is available from most dealers.

QUALITY-CONTROL PROCEDURES

Problems occur despite the manufacturer's best efforts in quality control of evacuated tubes. One should establish quality-control procedures to satisfy requirements for the evacuated tube. It is helpful to check a random few tubes from every lot number as they arrive in the stockroom and before they are placed in service. Stockroom people should be advised to send the phlebotomist one box of 100 tubes upon receipt of a new lot number. The tubes should then be tested as follows (Fig. 13-1).

1. Select several sample tubes at random from the box.
2. Visually inspect for additives and for foreign matter—that is, glass and stopper material.
3. Send the Microbiology Department two empty tubes to be checked for microbiologic contamination.

Quality Control On Vacuum Blood Drawing Tubes

Date	Lot No.
Lab	Tube Size
	Tube Color

Objectives: To establish quality control on all vacuum tubes as they are put into use.

Procedure: Each time a new lot number is put into use, tubes will be checked by designated laboratories: Venipuncture, Central Processing, Hematology, Metals, Microbiology, Chemistry.

A slip will accompany the tubes and will be checked indicating check to be done.

Each department will record results, initial, and send to ______________________

		Result	Initial			Result	Initial
	VENIPUNCTURE				METALS		
	Check for:				Check for:		
	Vacuum draw				Metals contamination		
	Visual inspection for						
	anticoagulants						
					MICROBIOLOGY		
					Check for:		
	CENTRAL PROCESSING				Sterility of tube		
	Check for:						
	Sample spin down						
	Stopper pull-off						
	Breakage:				CHEMISTRY		
	Cap off				Check for:		
	Spinning				Ammonia		
					Na, K		
	HEMATOLOGY						
	Check for:				Comments:		
	Clots						
	Strain anticoagulated						
	blood through gauze						
	Comments:						

Fig. 13-1. *Quality control on vacuum blood drawing tubes.*

4. Send Central Processing two tubes with blood to be checked for breakage during centrifugation and stopper removal.
5. Test four tubes for volume draw.
6. Send the Hematology Department two tubes (EDTA) with blood to be checked for clots.
7. Send the Metals Laboratory two empty tubes to be checked for metal contamination.
8. Send the Chemistry Laboratory two empty heparin tubes to be checked for ammonia contamination.

The test for vacuum draw has been written by the National Committee for Clinical Laboratory Standards (NCCLS).[1] When a container is tested for draw and fill accuracy at the time of its manufacture, the NCCLS standards require the draw to be within ±10% of the stated draw. At the time of the expiration date, the draw should be no less than 10% below the minimum volume allowable at the time of manufacture. For instance, at the time of manufacture, the allowable draw for a 10-ml tube should be anywhere from 9 ml to 11 ml. The allowable draw for that same tube at the date of expiration should be 8.1 ml (*see* Table 13-1).

Since this standard has been defined using conditions of 101 kPa and 20°C, corrections will be needed if other conditions are used during testing. A chart should be developed to facilitate these calculations. The method below is recommended by the NCCLS to test the vacuum draw. Materials needed to test capacity of tubes are

- one 50-ml precision bore certified buret, type 1, style 1, Class A, M.D.S.;
- 1-m length flexible vinyl or latex tubing connected to the buret tip;
- one 20-gauge, 1½-inch (38 mm) blood collection needle attached to the tubing; and
- deionized water.

The testing of 162 lots at our institution revealed different tube problems (*see table 13-2*).

The procedure itself comprises the following steps.

1. Fill buret with deionized water

***Table 13-1.** Tube Acceptability Ranges*

TUBE SIZE	10%	± RANGE
←	*ml*	→
15	1.5	−13.5 +16.5
10	1.0	− 9 +11
7	0.7	− 6.3 + 7.7
5	0.5	− 4.5 + 5.5

2. Bleed tubing and needle to remove air
3. Refill buret and bring meniscus to 0
4. Insert needle into stopper of tube
5. Open stopcock of buret, push needle through stopper of tube to be tested, and allow tube to fill completely
6. Read volume of water drawn to 0.1 after elevating tube so that it's meniscus is at the same height as the buret meniscus
7. Close the stopcock and remove the tube.
8. Refill and zero buret. Repeat test for each tube
9. Calculate results according to appropriate methods.

The acceptance criterion is that each tube be filled within ±10% of the labeled draw; otherwise, they will be deemed defective.

TUBE PROBLEMS

The most common problems found with evacuated tubes are long and short draws and metal contamination. The long or short draw in the clot tube is of little or no consequence as long as the required amount of serum to perform the test is obtained. A long or short draw in a tube that contains an anticoagulant can, in some instances, affect the test result, especially with the tubes used for hematology and coagulation studies.

In July 1981, a study was conducted in the Section of Laboratory Hematology at the Mayo Clinic to determine the effects of improper blood volume collections in the evacuated tubes for various hematologic measurements. Samples were collected from normal subjects for a complete blood count (CBC), sedimentation rate, activated partial thromoplastin time, prothrombin time, and differential leukocyte counts. The CBCs were collected into EDTA tubes designed to collect 4 ml of blood. The evacuated tubes were filled with 1 ml, 2 ml, 4 ml, and 5 ml of blood by syringe. The sedimentation rates were collected in citrate tubes designed to collect 2.4 ml of blood. The tubes were inoculated with 1.5 ml, 2.4 ml, and 4 ml of blood by syringe. The activated partial thromboplastin time (APTT) and prothrombin time (PT) were collected in tubes designed to collect 2.7 ml of blood. These samples were prepared with volumes of 1 ml, 2 ml, and 2.7 ml of blood with a syringe.

Peripheral blood smears for eye count differentials were prepared from the evacuated tubes designed to collect 4 ml of blood but containing 2 ml, 4 ml, and 5 ml of blood. Significant changes in the CBC occurred only in the 1-ml collection vials, but significant changes were observed in the PTs and APTTs if the proper volume were not collected. Eye count differentials demonstrated artifacts of red cell morphology, particularly in the 2-ml volume, as well as changes in the staining properties of the red cells.

False prolongation of partial thromboplastin time (APTT) and PT owing to inadequate filling of evacuated tubes has been reported by Humphreys and McPhedran.[2]

Microbiologic contamination may also be a problem.[3] Use of sterile evacuated tubes alleviates this problem by protecting the patient from the threat of contamination should there be some reflux of blood from the tube back into the patient's vein.

If nonsterile evacuated tubes are used for drawing a blood culture, hematology, or chemistry specimens, the culture media should be inoculated first, then the nonsterile tubes. False-positive blood culture results can occur because of backflow or reflux. Positive blood cultures were noted in Georgia and Massachusetts when the same syringe used to inoculate the evacuated tube that contained EDTA and that was contaminated with *Serratia* was used to inoculate the culture bottles.[4]

The risk of transmission of infection to patients during venipuncture because of the hazards of backflow is also possible. Katz and associates investigated the mechanism of backflow and the conditions under which it can occur and found it to be related to decreased venous pressure during blood collection.[5] Their study was prompted by a report by McLeish and associates that cited five cases of *Serratia bacteremia,* the causes of which were linked to the finding of *Serratia* in 3 of 13 evacuated tubes cultured.[6]

To avoid problems of false-positive blood cultures and possible infections of the patient, strict adherence to venipuncture technique is advisable and the use of tubes with sterile interiors is suggested. The Canadian government has made the use of the sterile evacuated tube mandatory.

The testing of 162 lots at our institution revealed different tube problems (*see* Table 13-2).

If a problem is found with a tube, several other tubes should be tested to

Table 13-2. *Tube Problems After Testing 162 Different Lots*

TUBE	PROBLEM	INCIDENCE	SOLUTION
		lots	
Clot	Calcium contamination	2	Unacceptable by laboratory
	Magnesium contamination	3	Acceptable by laboratory
	Zinc contamination	8	Two unacceptable for zinc determination
	Hemolysis	1	Replaced by supplier
	Cells collected around stopper during centrifugation	1	Acceptable when spun using glass beads
Heparin	Ammonia contamination	1	Not acceptable for ammonia determination
Citrate	Long draw	1	Replaced by supplier
EDTA	Short draw	4	One returned to supplier; three currently being used. Company is investigating
	Difficult stopper removal	1	Currently being used. Company is investigating

verify the finding. If, after further checking, a definite problem is noted, the factory representative should be contacted and asked to look into it. In our experience, the manufacturer has been extremely concerned and eager to correct the problem. Most often our entire lot will be replaced by the manufacturer at no cost.

By sample testing a few tubes each time a new lot number is received, one can avoid the problem of having to seek a substitute new lot number of tubes on very short notice. This can often be very difficult, if not impossible, to do. Occasionally, even after running quality-control tests, one may encounter a problem during use of the tubes. Long and short draws in tubes that contain additives should always be watched for.

These procedures, although they take time, will save time in the long run by reducing the number of redraws that have to be done. The department will save money because time and materials needed for redraws are costly. Most important, more accurate test results will be assured, thus improving patient care.

REFERENCES

1. National Committee for Clinical Laboratory Standards (NCCLS): Standard for Evacuated Tubes for Blood Specimen Collection, H1–A2, 2nd ed. Villanova, Pennsylvania, NCCLS, 1980
2. Humphreys RE, McPhedran P: False elevation of partial thromboplastin time and prothrombin time. JAMA 214:1702, 1970
3. Washington JA: The microbiology of evacuated blood collection tubes. Ann Intern Med 86:186, 1977
4. Center for Disease Control: False-positive blood cultures related to the use of evacuated nonsterile blood collection tubes—Georgia, Massachusetts. Morbid Mortal Weekly Rep 24:388, 1975
6. McLeish SW, Elder RH, Westwood J: Contaminated vacuum tubes. Can Med Assoc J 112:682, 1975

14

Managing the Phlebotomy Team

David Sperling

Mastering the techniques of phlebotomy is commendable, but managing the phlebotomy team may be even more rewarding for the right persons. But how does one put it all together? How does one make sure the employees are being treated fairly? How does one integrate the activities of the phlebotomy team into the long-term goals of the institution?

This chapter discusses some facets of management but does not cover all factors that relate to personnel management. In addressing your concerns, it is assumed that you are a person in middle management in your organization. You are supported by a personnel department that makes some institutional policies and that does some recruitment screening. You are responsible for some fiscal management in that you may take part in setting a budget, but you do not have final authority on integrating that budget into the overall fiscal management of your organization. You may or may not have final say on hiring or terminating an employee, but you do have input on what person to hire and whether that person should ever be terminated. In short, you have a boss (or several bosses) and are responsible for leading a team of phlebotomists. This discussion assumes that you have some latitude in recruiting your team and in scheduling their efforts, have input to their job descriptions, conduct their formal performance appraisals, and have input into their salary adjustments all within some policy framework in your organization. The discussion also assumes that you *want* to be a manager. Lacking that desire almost certainly means that you will not be as successful as you ultimately could be.

For purposes of this discussion, it is assumed that you work in an organization that has a formal personnel department. The initial screening of potential employees undoubtedly is accomplished by that department. When there are openings for positions, you are therefore sent some employment applications from which you will choose a small number for the formal interview process. The following is a brief review of a few of the *do's* of matters handled during the personal interview process.

DO put the person at ease.
DO lead them into conversation.
DO request that they talk about themselves. People usually become much more relaxed in a situation like this when they have ample time to tell about their background.
DO listen carefully to what is said. Describe the job carefully and seek their reaction to that description. Allow time for questions.

Here are a few *don'ts*.

DON'T take notes.
DON'T dominate the interview.
DON'T rush things.
DON'T ask questions that are illegal. (Seek the help of your personnel department in defining things.)

It is advantageous to have several candidates seeking the open position so that you can select the best one. In many organizations the salary offer ultimately is made by the personnel department, but *you* may wish to inform the candidate if new, important facts are discovered during the interview.

The following is a list of consensus of day-to-day supervision when a new person is recruited.

- Develop a standard way of introducing all new employees to their colleagues.
- Show them the entire work area and where the lounges are.
- Make sure they understand the basic responsibilities of the phlebotomy department.
- Make sure someone accompanies them on coffee breaks and to a lunch break.
- Sit down for 5 or 10 minutes with the new employee and the person responsible for training to develop a mutual understanding of the first few days' responsibilities.

For purposes of this discussion, it is assumed that you are responsible for a team that has 24-hour coverage. Obviously, you schedule most people during the busiest period (probably the early morning hours). A grid covering 24 hours of the day and 7 days of the week will be helpful in scheduling the employees. Use of part-time employees should be considered because scheduling may turn out to be one of the more difficult jobs because patients' needs must be satisfied and met according to the physician's orders.

A job description is essential in managing the phlebotomy team. This description summarizes the qualifications the team members need to do their job appropriately. It is very useful in planning and recruiting and it can be used to orient new employees to their responsibilities and duties and make them aware of the persons to whom they are responsible. It is a document used to develop performance standards and can be used for job evaluation. It must be a fair reflection of the job and usually should be no longer than two pages. Because the phlebotomist's responsibilities are dynamic, the job description should be audited periodically.

The job description itself should contain a title and a description of the job function, organizational relationships, duties, know-how, problem-solving, accountability, and any special factors. The function should be brief, usually a one-sentence description of how the job (task) fits into the organization in a broad way. The organizational relationship should describe to whom the person is responsible and from whom the person receives direction (supervision). The duties should describe a reasonably specific task list, including the frequency of performing the tasks. Documentation of a nonregular task (daily or weekly) but one intermittently requested on an individual basis can also be described. Action words, such as compiles, performs, lists, or checks, are words used to describe the duties. *Know-how,* broadly, is the minimum amount of formal education or training required to do the job. *Problem-solving* is a description of how much the duties follow rigidly established guidelines and procedures versus individual judgment and latitude. *Accountability* describes the amount and type of work that the person is responsible for. Special features can include rotations, special shifts, special strengths required, or if a driver's license is needed. The typical number of hours that a person works follows next.

The importance of phlebotomy work is being given more and more recognition. There are increasing pressures on professionalization. The job description should be used to recognize that increased professionalization. Job descriptions in the organization can be tailored to fit the institution's needs and, of course, can contain more than what is listed above.

Once a job description has been written, that description can be fitted into the organization's job classification. In most organizations, a job evaluation group looks at all descriptions and rates their relative importance to the organization for salary administration purposes. In some organizations, phlebotomists perform their work only on outpatients, whereas in others they deal strictly with inpatients. The very nature of inpatient work is more difficult because phlebotomists must go to the patient, must deal with sicker patients, and usually must have a higher level of knowledge about hospital practices, such as how to deal with patients with infectious disease. In most organizations, these differences allow for differentiation of the job classification for outpatient versus inpatient phlebotomists.

The best kind of performance appraisal can be done with the objective background of the job description, which allows for more objective performance ratings because the person doing the appraising has a benchmark against which judgments can be made. The formal performance appraisal should always be done in a face-to-face setting. Written documentation of objective measurements of performance is a must. The face-to-face interview process should encompass short- and long-range planning of expectations of both the supervisor and the employee. This is a great opportunity for positive strokes and written documentation of any possible outstanding performance. Above all, comments must be objective and must be understood fully by the employee. Regardless of the nature of the appraisal, some of it must be written, and it is a good idea to have the employee sign the documentation. It is also extremely important that both parties be honest with each other, particularly when subpar performance is part of the appraisal. Undocumented subpar performance, which later may become so unacceptable that termination is contemplated, leaves the supervisor with little objective evidence of previous poor performance. Termination then is extremely difficult because the evidence that may be needed to support deteriorating performance over time is missing.

On the basis of the performance appraisal, some sort of salary administration takes place. In virtually all organizations, every job has a minimum salary and a maximum salary. It is between these two extremes that a supervisor must work. From the performance appraisal, the supervisor will recommend either less than a usual increase, a typical increase, or an above-average increase. Most organizations recognize that most persons fall into a "usual category" for salary administration purposes. Salary adjustment that is given should be a fair reward for the work performed. In many organizations, persons recruited at the minimum salary for that classification advance to the maximum for that class in about 5 years. In many organizations, allowance is made for special duty pay when certain special factors are considered. If 80% to 90% of personnel in a particular department basically do what is covered in the job description but some do extra duties, there can be some acknowledgement of this in the salary for that particular job. Also, in any organization, frequently a few persons are simply outstanding but do not wish to progress to more responsible positions. Because they are so valuable and do such consistently outstanding work, they can be rewarded with special merit pay, which is an allowance over the usual maximum for that particular job. Finally, most organizations recognize off-hours work (any schedule other than the normal Monday through Friday, 8:00 a.m. to 5:00 p.m. workweek), with extra pay sometimes being defined as shift differentials.

This review has been brief, but numerous books have explored in depth the particular topics discussed. None, however, can assure you that you will love working with people. It is worth repeating that to be a good manager you must first *want* the job. You must also clearly understand that this involves being genuinely concerned about people and all the challenges and opportunities employees present to you.

15

Summary of Considerations in the Collection of Specimens

Marjorie Gamm
Ruth Mangan
Jean M. Slockbower

Factors such as staffing, supervision, communication, and environment influence the success of a specimen collection team. Patterns of collection vary according to the type of service. An outpatient specimen collection service will have defined needs that differ in some respects from a hospital specimen collection service, just as a small hospital's needs will differ from a large hospital's needs.

Nowhere are good systems and management more important than in a collection service. Today the cost of obtaining the patient's specimen is often more expensive than the cost of routine laboratory tests. It is essential to determine patterns of service and staffing needs, correlate productivity indices, and ascertain the cost of these laboratory services.

The following materials relate to phlebotomy service but could be adjusted to other collection services.

OUTPATIENT PHLEBOTOMY

The number of patients that can be served in any given time in an outpatient setting should be determined. Obviously the morning load demand will be greater than the afternoon because of patient instructions on fasting for many laboratory tests.

Quotas per time intervals, queue lines, and waiting times should be established and audited periodically. A reasonable goal would be to see 95% of scheduled patients within 30 minutes.

Patient details that need to be checked include patient identification, instructions on fasting or a particular type of meal, and test requisition request forms. Patient care should be attended to because there are always patients who are worried or those having problems. Special attention should be given to the patient who becomes ill, and a medical resident should be on-call to see these patients. A team of phlebotomists should be trained in cardiopulmonary resuscitation techniques.

PHLEBOTOMY IN THE SMALL LABORATORY

"Organization" is the key word for the technologist who handles phlebotomy in the small hospital laboratory. This technologist will be performing phlebotomy as only a small part of the total workload. The workday may comprise procedures in blood banking, chemistry, urinalysis, microbiology, hematology, x-rays, ECGs, and pulmonary function testing. The phlebotomy area must be located near the laboratory in case the technologist needs to adjust the laboratory testing workflow to accommodate the venipuncture interruption.

Each technologist should have his own phlebotomy equipment and assume responsibility for it. It should be readily available and organized by each person to his own particular needs, thus ensuring that he is well equipped and ready to do phlebotomy at any time.

Adequate scheduling is necessary to good organization in the phlebotomy area. The schedule should be detailed, fair to all employees, and rotated among the entire staff.

REQUISITIONS

In a hospital laboratory, one of the first early morning tasks is to review the requisitions for duplicate orders. The requisition slip originates at the nurses' station. In addition to the patient's name, identification number and location, it lists all the tests available in the laboratory. The tests ordered on the patient are indicated by a checkmark, as well as the time of day that the tests are to be drawn. This requisition may be a multipart form comprising one copy for billing purposes and a report copy for the medical records department, the physician, the laboratory files, and the patient's chart. The requisitions should be checked against the hospital census sheet to make sure that the patients have not been transferred or released since the laboratory work was ordered. The requisitions should then be divided by location within the facility for efficiency of collection.

SCHEDULING

One person should be designated to arrive at the laboratory 15 to 30 minutes early in the morning to handle the requisitions and to check for any special requests, such as early surgicals or baby bilirubin levels. There should be a prepared schedule assigning each technologist to a specific area for 1 week at a time for these early draws. These areas should include obstetrics, the nursery, pediatrics, medical, intensive care, and surgical. The assignments should be rotated so that the same technologists do not have difficult or undesirable areas each time (*see* Chart 15-1).

Additional scheduling is needed throughout the day to ensure that one person will assume responsibility for additional draws for such things as "stats," outpatients, and "now" draws. One person could be assigned to handle "stats," with another person as first backup in case of changes in the workload (*see* Chart 15-2).

Another example of scheduling would be in the promotion of additional drawing times throughout the day. This could diminish the number of "now" draws and could result in more batching of orders and testing (*see* Chart 15-3).

TRAINING AND EVALUATION

New employees may come to the laboratory at various levels of phlebotomy skill development. An initial evaluation should be made with each new employee to establish the level of competency. Areas to be evaluated include location of major vessels, difficult draws, use of various evacuated tubes, unusual procedures such as blood alcohol tests and arterial punctures, communication with nursing personnel, and preserving the integrity of the specimen. The procedures with which a new employee feels least confident should be noted and training implemented. This list should be reviewed at 3-week intervals until the technologist feels comfortable in all situations. The procedures discussed during the new technologist's evaluation should also be written in a phlebotomy procedure manual.

This schedule is used for early morning draws.

Chart 15-1. *Examples of Schedules Used at One Hospital for Phlebotomy Technologists*

OB NRSY ICU
SURGICAL
PEDIATRICS — DATE ________________
MEDICAL I — IN THE CASE OF ANY ABSENT INDIVIDUAL, SLIPS SHOULD BE DIVIDED
MEDICAL II — ACCORDINGLY AND TECHS WILL MOVE TO THE NEXT VACANCY.
C&R I & II

DAY	OB NRSY & ICU	SURGICAL	PEDS	MEDICAL I	MEDICAL II	C&R I AND II
Monday						
Tuesday						

This schedule is used for additional draws throughout the day.

Chart 15-2. *Collecting Schedule*

Date ________________

	10:30 a.m.	1:30 p.m.	3:30 p.m.
Monday			
Tuesday			

This schedule is for STAT collections. Note the additional back-up columns that are available if the assigned technologist is unavailable.

Chart 15-3. *Stat Collection*

	ASSIGNED	FIRST BACK UP	SECOND BACK UP
Monday			
Tuesday			

PHLEBOTOMY IN THE LARGE HOSPITAL

Personnel should be scheduled to accommodate the workload of the 24-hour day by determining the amount and distribution of work, including routine and priority orders on each shift. Then by developing workload recording data, one can determine total time for priority and routine work. With these data, one can assign staff to a work rotation in accordance to volume and time to do the work on each shift.

Workload recording measurements made in our phlebotomy sections used a standardized time study format from the College of American Pathologists. The unit value determined by the time study represents a measure of the personnel time required to perform a laboratory test or service. One unit is equivalent to 1 minute of technical or clerical time. Examples of the results from studies done in 1980 for outpatient and hospital phlebotomy service are presented in Table 15-1. The accuracy of the unit value depends on complete documentation of all the steps involved to perform the service and the correct measurement of the time to perform each activity.

Table 15-1. *Workload Recording*

	TIME/MINUTE	UNIT TIME/ONE PATIENT
Outpatient service		
Check in at receptionist	0.53	
Rooming patient	0.40	
Label and tube assemble	1.08	
Adult specimen collection		
Chair	3.29	5.27
Bed	3.42	5.44
Wheelchair	4.58	6.60
Pediatric specimen collection (two technicians)	8.80	10.82
Hospital service		
Routine order		
Initial handling of order	2.40	
Travel time	0.97	
Adult specimen collection	4.84	
Verify and dispatch	0.47	8.68

These data may be used to provide a productivity index for management purposes, important because one needs to run this type of service efficiently and as economically as possible. Cost analysis can be done by analyzing the timing measurements for each function performed; identifying the consumable supplies and equipment, other miscellaneous expenses for the services, and the space allocated to the service; and auditing the personnel list to obtain total full-time equivalents.

Until the time there is only noninvasive monitoring of patient analytes, specimen collection will remain an important aspect of the clinical laboratory. Therefore, all specimen-collection personnel must be well trained and provided with reliable procedures, quality supplies, and responsible management. Only then can we ensure that the laboratory is testing a biologically representative patient specimen and, in turn, supporting the patient-care physician by providing reliable information on the diagnosis, prevention, or treatment of disease.

Index

Numbers followed by an "f" indicate a figure or chart; "t" following a page number indicates tabular material.